Lecture Notes in Mathematics

A collection of informal reports and seminars
Edited by A. Dold, Heidelberg and B. Eckmann, Zürich

T0240469

135

Wilhelm Stoll

University of Notre Dame, Notre Dame/IND/USA

Value Distribution of Holomorphic Maps into Compact Complex Manifolds

Springer-Verlag
Berlin · Heidelberg · New York 1970

This work is subject to copyright. All rights are reserved, whether the whole or part of the material is concerned, specifically those of translation, reprinting, re-use of illustrations, broadcasting, reproduction by photocopying machine or similar means, and storage in data banks.

Under § 54 of the German Copyright Law where copies are made for other than private use, a fee is payable to the publisher, the amount of the fee to be determined by agreement with the publisher.

© by Springer-Verlag Berlin · Heidelberg 19 0 Library of Congress Catalog Card Number 75-121987 Printed in Germany.
Title No. 3291

PREFACE

The theory of value distribution in several complex variables received a new impetus by the results of Levine [14] and Chern [2] in 1960. In 1965, Bott and Chern [1] developed a theory of equidistribution of zeros of holomorphic sections in vector bundles. In 1967, a first main theorem of value distribution for holomorphic maps into the projective space was given in [28]. In 1968, Hirschfelder [6] was able to extend this theory to holomorphic maps into compact complex manifolds for admissible families of analytic sets parameterized by a homogeneous Kaehler manifold. Independently, at the same time, Wu [33] developed a similar theory, which treated only holomorphic maps of fiber dimension 0 into a compact Kaehler manifold for the point family.

During the Fall Quarter of 1969, the author conducted a research seminar at Stanford University, where he presented the theory, as given in these Lecture Notes. The proximity form was now constructed directly and explicitly. This made it possible to drop the homogeneity condition of Hirschfelder's approach.

The author received great help by a communication of Hirschfelder showing that Wu's proximity form is also a proximity form in the case of positive fiber dimension (Hirschfelder [7a]). This communication was received at the beginning of the seminar. Hirschfelder's observation helped the author to define the singular potential (Definition 5.1) and to show that it is a proximity form.

The theory as presented here, owes much to Hirschfelder and Wu. It gives new results as well as it represents and unifies ideas and results of Chern, Hirschfelder, Wu and the author. Proofs within the theory are given whether new or not. Outside results of other theories, as Hodge theory on Kaehler manifolds, multiplicity of holomorphic maps and the continuity of the fiber integral are used without proofs.

An exception is made with the theory of integration over the fibers of a differential or holomorphic map. The author learned of this operator from Bott and Chern [1]. The use of the operator seems to be spreading. However, no account with precise statements and complete proofs seems to have appeared. Therefore, with the encouragement of A. Andreotti and S. S. Chern, the author has given such an account in Appendix II without any claim to originality.

Appendix I contains a group of highly technical results requiring complicated proofs. A large part of this appendix consists of an almost literal reproduction of parts of §2 of Hirschfelder's thesis [6]. This is not easily accessible since much of it was suppressed in [7] due to space restrictions.

Originally, it was anticipated to include in these notes an outline of the theory of Bott and Chern [1], and to show, how their equidistribution theory in ample vector bundles can be obtained from the theory here respectively from [28]. However, this will appear at another place.

This work was done at Stanford University while the author was supported by the University of Notre Dame and by the National Science Foundation under Grant NSF GP7265. The author wishes to express his gratitude to these institutions for their help and support to make this work possible.

Wilhelm Stoll
Notre Dame and Stanford
Spring 1969

CONTENT

§1. Introduction

At first, a short outline of the first main theorem for meromorphic functions on the complex plane shall be given.[1)] With the exception of one term, the "deficit", it contains all the ingrediences of the theory in several variables and, therefore, may facilitate the understanding of the general theory.

On any complex manifold, the exterior derivative d splits into a complex part ∂ and its conjugate part $\bar{\partial}$ such that $d = \partial + \bar{\partial}$. Define

$$(1.1) \qquad d^{\perp} = i(\partial + \bar{\partial}) = -d^{c}$$

Let $\mathbb{P}$ be the Riemann sphere. If $\mathbb{P}$ is realized as a sphere of diameter 1 in $\mathbb{R}^3$, let $\|w:a\|$ be the euclidean distance in $\mathbb{R}^3$ from $a \in \mathbb{P}$ to $w \in \mathbb{P}$. Regard a as fixed and w as variable. On $\mathbb{P}$, a Kaehler metric exists, whose exterior form of bidegree (1,1) is given by

$$\omega = \frac{1}{\pi} dd^{\perp} \log \|w:a\|$$

on $\mathbb{P} - \{a\}$ for every $a \in P$. Of course ω is independent of a. For $0 < r < \infty$ define

$$G_r = \{z \in \mathbb{C} \mid |z| < r\}$$
$$\Gamma_r = \{z \in \mathbb{C} \mid |z| = r\}.$$

For $0 < r_0 < r$, define

$$\psi_r(z) = \begin{cases} 0 & \text{if } z \notin G_r \\ \log \frac{r}{|z|} & \text{if } z \in G_r - G_{r_0} \\ \log \frac{r}{r_0} & \text{if } z \in G_{r_0} \end{cases}$$

Let $f\colon \mathbb{C} \to \mathbb{P}$ be a non-constant, holomorphic map. For each $a \in \mathbb{P}$ and $z \in \mathbb{C}$, the a-multiplicity $\nu_f^a(z) \geqq 0$ of f at z is defined. Denote

$$n_f(r,a) = \sum_{z \in \bar{G}_r} \nu_f^a(z) \geqq 0$$

$$N_f(r,a) = \sum_{z \in G_r} \psi_r(z)\nu_f^a(z) = \int_{r_0}^{r} n_f(t,a)\,\frac{dt}{t}$$

$$A_f(t) = \int_{G_t} f^*(\omega) \geqq 0$$

$$T_f(r) = \int_{G_r} \psi_r f^*(\omega) = \int_{r}^{r_0} A_f(t)\,\frac{dt}{t} \geqq 0$$

$$m_f(r,a) = \frac{1}{2\pi}\int_{\Gamma_r} \log \frac{1}{\|f:a\|}\, d^{\perp}\psi_r \geqq 0$$

where f^* is the pullback of forms from $\mathbb{P}$ to $\mathbb{C}$. Names are : $n_f(r,a)$ counting function, $N_f(r,a)$ integrated counting function, $A_f(t)$ spherical image, $T_f(r)$ characteristic function, $m_f(r,a)$ proximity function. Moreover, $N_f(r,a)$, $A_f(r)$, $T_f(r)$, $m_f(r,a)$ are non-negative, continuous functions of r and a, and

$$T_f(r) \to \infty \qquad \text{for } r \to \infty$$

The <u>first main theorem</u> holds for $r > r_0$:

$$T_f(r) = N_f(r,a) + m_f(r,a) - m_f(r_0,a).$$

Let $C^0(P)$ be the algebra of continuous functions on $\mathbb{P}$. Let $L\colon C^0(P) \to R$ be a linear, increasing map, i.e., $L(h_1) \leqq L(h_2)$ if $h_1 \leqq h_2$. Then $L(1) \geqq 0$. The <u>defect of f for L</u> is given by

$$0 \leqq \delta_f(L) = \underline{\lim}_{r \to \infty} \frac{L(m_f(r,a))}{T_f(r)}$$

$$= L(1) - \overline{\lim}_{r \to \infty} \frac{L(N_f(r,a))}{T_f(r)} \leqq L(1)$$

If L_1 and L_2 are two such maps, then also $L_1 + L_2$ is linear and increasing and

$$\delta_f(L_1 + L_2) \geqq \delta_f(L_1) + \delta_f(L_2)$$

Now, several examples of operators L are helpful.

<u>1. Example.</u> Average $\hat{L}$: For $h \in C^0(\mathbb{P})$ define

$$\hat{L}(h) = \int_{\mathbb{P}} h\omega$$

Then $\hat{L}(1) = 1$. The <u>averaging formula</u> holds

$$T_f(r) = \hat{L}(N_f(r,a)).$$

2. Example. Image operator L_f. For $h \in C^0(\mathbb{P})$ define

$$L_f(h) = \int_{f(\mathbb{C})} h\omega$$

Then $b = L_f(1)$ is the measure of the image set normalized such that $\mathbb{P}$ has measure 1. Hence $0 \leqq b \leqq 1$. Because $N_f(r,a) = 0$ for $a \in \mathbb{P} - f(\mathbb{C})$,

$$T_f(r) = \hat{L}(N_f(r,a)) = L_f(N_f(r,a))$$

Hence, $0 \leqq \delta_f(L_f) = b - 1$, which implies $b = 1$; meaning that f assumes almost every value.

3. Example. Dirac operator L_A. Let $A \subseteq \mathbb{P}$ be a finite subset. For $h \in C^0(\mathbb{P})$ define

$$L_A(h) = \sum_{a \in A} h(a)$$

Abbreviate $L_a = L_{\{a\}}$ and $\delta_f(a) = \delta_f(L_a)$. Then $\delta_f(a)$ is the Nevanlinna defect.. Because $L_A = \sum_{a \in A} L_a$ the inequality $\sum_{a \in A} \delta_f(a) \leqq \delta_f(A)$ is true. The second main theorem implies the defect relation

$$\sum_{a \in A} \delta_f(a) \leqq \delta_f(A) \leqq 2 .$$

If $a \notin f(\mathbb{C})$, then $\delta_f(a) = L_a(1) = 1$. Hence $\mathbb{P} - f(\mathbb{C})$ contains at most two points.

In several variables, value distribution is concerned with the following problem: Let $f: M \to N$ be a holomorphic map of a non-compact complex manifold[2)] M of dimension m into a complex manifold N of dimension n. Let $\mathfrak{A} = \{S_a\}_{a \in A}$ be a family of p-dimensional analytic subsets of N. When is $f^{-1}(S_a) \neq \emptyset$ for almost all a?

Some assumptions have to be made, to expect reasonable results. So, A is a Kaehler manifold and $\mathfrak{A}$ is an admissible family as defined in §2. Let $s = n - p > 0$ be the codimension of S_a. Then f is said to be adapted to $\mathfrak{A}$ at $z \in M$ for $a \in A$, if and only if open neighborhoods U of z and V of a exist such that $f^{-1}(S_t) \cap U$ is either empty or has pure codimension s. Define $q = m - s$. For the sake of simplicity, it shall be assumed in this introduction alone that f is adapted to $\mathfrak{A}$ at every $z \in M$ for every $a \in A$. If $(z,a) \in M \times A$, an intersection multiplicity $\nu_f^a(z) \geqq 0$ is defined in §2, such that $f^{-1}(S_a) = \operatorname{supp} \nu_f^a$.

If A is a connected Kaehler manifold, and if $\mathfrak{A} = \{S_a\}a \in A$ is an admissible family, a non-negative form Ω of class C^∞ and bidegree (s,s) on N and, for every $a \in A$, a non-negative form Λ_a of class C^∞ and of bidegree (s-1,s-1) on $N - S_a$ are constructed such that $d^\perp d\Lambda_a = \Omega$ on $N - S_a$. Here Λ_a is singular on S_a, and depends "nicely" on $a \in A$.

On M, suppose that a non-negative differential form χ of class C^1 and of bidegree (q,q) is given such that $d\chi = 0$. Suppose that χ is positive on some non-empty open subset of M. A <u>bump</u> $B = (G,\Gamma,g,\gamma,\psi)$ on M is a collection consisting of open, non-empty, relative compact subsets G and g of M with C^∞ boundaries $\Gamma = \overline{G} - G$ and $\gamma = \overline{g} - g$ oriented to the exterior of G respectively g, and of a non-negative, continuous function $\psi: M \to \mathbb{R}$, such that $0 \leqq \psi \leqq R$ on M, such that $\psi | g = R$ and $\psi | (M - G) = 0$ are constant and such that $\psi | (\overline{G} - g)$ is of class C^2. Define $G(r) = \{z \in M | \psi(z) > R - r\}$.[3)]

Then the following integrals exist and have non-negative integrands

$$n_f(G,a) = \int_G \nu_f^a \chi$$

$$N_f(G,a) = \int_G \psi \nu_f^a \chi = \int_0^R n_f(G(t),a)\, dt$$

$$A_f(G) = \int_G f^*(\Omega) \wedge \chi$$

$$T_f(G) = \int_G \psi f^*(\Omega) \wedge \chi = \int_0^R A_f(G(t))\, dt$$

$$m_f(\Gamma,a) = \int_\Gamma f^*(\Lambda_a) \wedge d^\perp \psi \wedge \chi$$

$$m_f(\gamma,a) = \int_\gamma f^*(\Lambda_a) \wedge d^\perp \psi \wedge \chi\, .$$

Also the following integral exists, but nothing about the sign of its integrand can be said:

$$D_f(G,a) = \int_{G-g} f^*(\Lambda_a) \wedge dd^\perp \psi \wedge \chi .$$

All the functions are continuous in $a \in A$. They are named: $n_f(G,a)$ counting function, $N_f(G,a)$ integrated counting function, $A_f(G)$ spherical image, $T_f(G)$ characteristic function, $m_f(\Gamma,a)$ proximity function, $m_f(\gamma,a)$ proximity remainder, $D_f(G,a)$ deficit.

The first main theorem holds

$$T_f(G) = N_f(G,a) + m_f(\Gamma,a) - m_f(\gamma,a) - D_f(G,a).$$

Let $C^0(A)$ be the algebra of continuous functions on A. Let ω be the volume element of the Kaehler manifold A normalized such that $\int_A \omega = 1$. The average $\hat{L}: C^0(A) \to \mathbb{R}$ is defined by

$$\hat{L}(h) = \int_A h\omega \qquad \text{for} \qquad h \in C^0(A).$$

Then the <u>averaging formula</u>

$$T_f(G) = \hat{L}(N_f(G,a))$$

holds. Moreover, a continuous non-negative form $\hat{\Lambda}$ of bidegree $(s-1,s-1)$ can be constructed on N such that

$$\mu_f(\Gamma) = \hat{L}(m_f(\Gamma,a)) = \int_\Gamma f^*(\hat{\Lambda})\wedge d^\perp\psi\wedge\chi \geqq 0$$

$$\mu_f(\gamma) = \hat{L}(m_f(\gamma,a)) = \int_\gamma f^*(\hat{\Lambda})\wedge d^\perp\psi\wedge\chi \geqq 0$$

$$\Delta_f(G) = \hat{L}(D_f(G,a)) = \int_{G-\bar{g}} f^*(\hat{\Lambda})\wedge dd^\perp\psi\wedge\chi$$

Then

$$\Delta_f(G) = \mu_f(\Gamma) - \mu_f(\gamma).$$

Now, consider the <u>image operator</u> L_f. Define $J_f = \{a \in A \mid f^{-1}(S_a) \neq \phi\}$. For $h \in C^0(A)$ define

$$L_f(h) = \int_{J_f} h\omega$$

Then $0 \leqq b_f = L_f(1) \leqq 1$ and b_f is the measure of J_f. Obviously, $b_f = 1$, if and only if $f(M)$ intersects S_a for almost every $a \in A$. Since $N_f(G,a) = 0$ if $a \in A - J_f$,

$$T_f(G) = \hat{L}(N_f(G,a)) = L_f(N_f(G,a)).$$

Now $m_f(\gamma,a) \geqq 0$ implies $L_f(m_f(\gamma,a)) \leqq \mu_f(\gamma)$. Hence the first main theorem implies

$$b_f T_f(G) = T_f(G) + L_f(m_f(\Gamma,a)) - m_f(\gamma,a) - D_f(G,a))$$

or

$$0 \leqq (1 - b_f)T_f(G) \leqq L_f(D_f(G,a)) + \mu_f(\gamma).$$

Now, M shall be exhausted by bumps. Let I be a directed set, then a net $\mathfrak{B} = \{B_r\}_{r\in I}$ of bumps $B_r = (G_r,\Gamma_r,g,\gamma,\psi_r)$ is said to <u>exhaust</u> M, if and only if g and γ do not depend on r and for every compact subset K of M an element $r_0(K) \in I$ exists such that $\psi_r(z) > 0$ if $z \in K$ and $r \geqq r_0(K)$ (especially $G_r \supset K$ if $r \geqq r_0(K)$). Then an element $r_1 \in I$ exists such that $T_f(G_r) > 0$ if $r \geqq r_1$. Define

$$\Delta_f^0(\mathcal{L}) = \lim_{\mathcal{L}} \sup \frac{L_f(D_f(G_r, a))}{T_f(G_r)}$$

$$\Delta_f(\mathcal{L}) = \lim_{\mathcal{L}} \sup \frac{\Delta_f(G_r)}{T_f(G_r)}$$

$$\mu_f(\mathcal{L}) = \lim_{\mathcal{L}} \sup \frac{\mu_f(\gamma)}{T_f(G_r)}$$

Then

$$0 \leqq 1 - b_f \leqq \Delta_f^0(\mathcal{L}) + \mu_f(\mathcal{L}).$$

The <u>defects</u> $\Delta_f^0(\mathcal{L})$ and $\mu_f(\mathcal{L})$ can be computed or estimated in several cases:

<u>1. Divisor case:</u> $s = m - q = n - p = 1$. Then $\mathfrak{A}$ is a family of <u>divisors</u> on N. Assume that χ is positive on all of M. For g take an open, non-empty, relative compact subset of M whose boundary $\gamma = \overline{g} - g$ is a C^∞-boundary manifold of g. Let I be the set of all open, relative compact subsets G of M with $G \supset \overline{g}$ such that $\Gamma = \overline{G} - G$ is a C^∞-boundary manifold of G. For $G \in I$, define a continuous function ψ_G uniquely by the conditions:

1. $\psi_G | \overline{g} = R(G) > 0$ and $\psi_G | (M - G) = 0$ are constant.
2. On $\overline{G} - g$, the function ψ_G is of class C^∞ and

$$dd^\perp \psi_G \wedge \chi = 0 \text{ on } G - \overline{g}.$$

3. It is

$$\int_\Gamma d^\perp \psi_G \wedge \chi = \int_\gamma d^\perp \psi_G \wedge \chi = 1.$$

Hence, ψ_G is the solution of the Dirichlet problem of an elliptic differential equation where the boundary value 0 is prescribed on γ and the solution is constant and positive on the boundary part Γ. By Stokes theorem, the integrals in 3. are equal. They are positive. Hence $R(G)$ can be chosen, such that these integrals are 1. If $R(G) \to \infty$ for $G \Leftrightarrow I$, then $T_f(G) \to \infty$ for $G \Leftrightarrow I$. Because $\hat{\Lambda}$ has bidegree $(0,0)$ on N, it is a continuous function on the compact manifold N. Hence $\hat{\Lambda} \leqq C$ for some positive constant C. Therefore $\mu_f(\gamma) \leqq C$. Obviously, $D_f(G,a) = 0$, hence $\Delta_f^0(\mathcal{G}) = \Delta_f(\mathcal{G}) = 0$. This implies the following

Theorem: If, in this divisor case, $T_f(G) \to \infty$ for $G \Leftrightarrow I$, then $b_f = 1$, meaning that $f(M)$ intersects S_a for almost every $a \in A$.

In the other cases, the exhaustion family $\mathcal{G}$ is defined by an exhaustion function. A non-negative function $h: M \to \mathbb{R}$ of class C^∞ is said to be an exhaustion function if and only if for every $r > 0$ the set

$$G_r = \{z \in M \mid h(z) < r\}$$

is non-empty, relative compact and open. Define

$$\Gamma_r = \{z \in M \mid h(z) = r\}.$$

Take $r_0 > 0$, such that $dh = 0$ on Γ_{r_0}. Then $\gamma = \Gamma_{r_0}$ is a C^∞-boundary manifold of $g = G_{r_0}$. Let I be the set of all $r > r_0$ such that $dh \neq 0$ on Γ_r, which is a set of almost all $r > r_0$. For $r > r_0$, define ψ_r by $\psi_r(z) = 0$ if $z \in M - G_r$, $\psi_r(z) = r - h(z)$ if $z \in G_r - g$ and

$\psi_r(z) = r - r_0$ if $z \in g$. If $r \in I$, then $B_r = (G_r, \Gamma_r, g, \gamma, \psi_r)$ is a bump and $\mathcal{B}_h = \{B_r\}_{r \in I}$ exhausts M. Write $T_f(r) = T_f(G_r)$ $N_f(r,a) = N_f(G_r,a)$ etc. Then

$$T_f(r) = \int_{r_0}^{r} A_f(t)\,dt \qquad N_f(r,a) = \int_{r_0}^{r} n_f(t,a)\,dt$$

Because $d^\perp\psi_r = -d^\perp h$ is independent of r on γ, the proximity remainder $m_f(r_0,a)$ and its average $\mu_f(r_0)$ do not depend on $r \in I$. Moreover, $T_f(r) \to \infty$ for $r \to \infty$. Hence $\mu_f(\mathcal{B}_h) = 0$.

2. Pseudo-concave case. The connected, noncompact manifold M is said to be pseudoconcave, if and only if an exhaustion function h exists, such that its Levi form $d^\perp dh$ is nonpositive outside a compact set. Take such a function h and take r_0 so large that $d^\perp dh \leqq 0$ on M - g. Then $D_f(G_r,a) \leqq 0$. Hence $\Delta_f^0(\mathcal{B}_h) \leqq 0$.

Theorem. If M is pseudoconcave, then f(M) intersects S_a for almost all $a \in A$.

3. Pseudo-convex case. The connected, non-compact complex manifold M is called pseudoconvex if and only if an exhaustion function h exists such that its Levi form $d^\perp dh$ is non-negative outside a compact set. Take such an exhaustion function h and take r_0 so large that $d^\perp dh \geqq 0$ on M - g. Then

$$D_f(r,a) = \int_{G_r} f^*(\Lambda_a) \wedge d^\perp dh \wedge \chi \geqq 0.$$

and

$$L_f(D_f(r,a)) \leq \hat{L}(D_f(r,a)) = \Delta_f(r)$$

which implies $\Delta_f^0(\mathcal{B}_h) \leq \Delta_f(\mathcal{B}_h)$.

Theorem. If M is pseudoconvex, if h is a pseudoconvex exhaustion, and if

$$\frac{\Delta_f(r)}{T_f(r)} \to 0 \quad \text{for} \quad r \to \infty$$

then f(M) intersects S_a for almost every $a \in A$.

Observe that every Stein manifold is pseudoconvex. On every Stein manifold (and only on these) an exhaustion function $h: M \to \mathbb{R}$ exists such that $d^\perp dh > 0$ on all of M. Then $d^\perp dh$ is the exterior (1,1)-form associated to a Kaehler metric on M. Now, a natural choice of χ can be made:

$$\chi = d^\perp dh \wedge \dots \wedge d^\perp dh \qquad \text{(q-times)}$$

Therefore, on Stein manifolds, the theory depends only on the choice of the exhaustion function h and all terms can be expressed in terms of h and α.

§2. Admissible Families

Let N be a complex manifold of dimension n. A family $\mathfrak{A} = \{S_a\}$ is said to be admissible[4] if and only if there exists a triplet $N \xleftarrow{\tau} F \xrightarrow{\pi} A$, called a defining triplet such that

1) Both F and A are complex manifolds.

2) The maps $\tau: F \to N$ and $\pi: F \to A$ are proper, surjective, holomorphic and regular.[5]

3) If $a \in A$, then $S_a = \tau\pi^{-1}(a) \neq N$. The restriction $\tau: \pi^{-1}(a) \longrightarrow S_a$ is injective (and hence bijective).

Because τ is proper, surjective, holomorphic and regular $\pi^{-1}(a)$ is a compact, p-dimensional, smooth[5] complex submanifold of F with $p = \dim F - \dim A$. Because τ is proper and because $\tau: \pi^{-1}(a) \to S_a$ is bijective, S_a is a compact, pure p-dimensional analytic subset of N. Moreover $0 \leqq p < n$, because $S_a \neq N$; hence, S_a is nowhere dense in N. If S_a consists of simple points only, then $\tau: \pi^{-1}(a) \to S_a$ is biholomorphic. Define $s = n - p$ as the codimension of $\mathfrak{A}$.

The maps $\pi: F \to A$ and $\tau: F \to N$ define differentiable fiber bundles, but not necessarily holomorphic fiber bundles.

Now several examples of admissible families shall be given:

1. Example: The point family.

Let N be a complex manifold of dimension n. Define $\mathfrak{A}_N = \{\{x\}\}_{x \in N}$ as the family of points of N. Define A = N and $F = \{(x,x) \mid x \in N\} \subseteq N \times N$. Let $\tau: F \to N$ and $\pi: F \to N$ be the natural projections. Then $N \xleftarrow{\tau} F \xrightarrow{\pi} N$ is a defining triple for N.

For the other examples, a complex vector space V of dimension $v + 1$ with $0 < v < \infty$ is used. Let $\mathbb{P}(V)$ be the associated projective space. Let $\mathbb{P}: V - \{0\} \to \mathbb{P}(V)$ be the natural projection such that $P(\mathfrak{z}) = P(\mathfrak{w})$ if and only if $\mathfrak{z} = \lambda\mathfrak{w}$ for some $\lambda \in \mathbb{C} - \{0\}$. The same letter $\mathbb{P}$ is used for all vector spaces. Denote $V[p] = V \wedge \dots \wedge V$ (p-times). If $0 \leq p \leq v$, define the <u>Grassmann cone</u> by

$$\tilde{G}_p(V) = \{\mathfrak{v}_0 \wedge \dots \wedge \mathfrak{v}_p | \mathfrak{v}_\mu \in V\} \subseteq V[p+1].$$

The <u>Grassmann manifold</u> $G_p(V) = \mathbb{P}(\tilde{G}_p(V) - \{0\})$ is a smooth, compact, connected, complex submanifold of $\mathbb{P}(V[p+1])$ and has dimension $(p+1)(v-p)$. For $0 \neq \mathfrak{v} \in \tilde{G}_p(V)$, the $(p+1)$-dimensional linear subspace $E(\mathfrak{v}) = \{\mathfrak{z} \in V | \mathfrak{z} \wedge \mathfrak{v} = 0\}$ is defined. If $\mathfrak{v} = \mathfrak{v}_0 \wedge \dots \wedge \mathfrak{v}_p$, then $E(\mathfrak{v}) = \mathbb{C}\mathfrak{v}_0 \oplus \dots \oplus \mathbb{C}\mathfrak{v}_p$. If $a \in G_p(V)$, then $E(a) = E(\mathfrak{v})$ is well defined by $\mathfrak{v} \in \mathbb{P}^{-1}(a)$. Moreover, E maps $G_p(V)$ bijectively onto the set of all $(p+1)$-dimensional linear subspaces of V. If $a \in G_p(V)$, define

$$\ddot{E}(a) = \mathbb{P}(E(a)) = \mathbb{P}(E(a) - \{0\}) \subseteq \mathbb{P}(V)$$

Then $\ddot{E}$ maps $G_p(V)$ bijectively onto the set of all p-dimensional, projective linear subspaces of $\mathbb{P}(V)$. Obviously, $G_0(V) = \mathbb{P}(V)$.

<u>2. Example</u>.

Let V be a complex vector space of dimension $v + 1$ with $0 < v < \infty$. For $0 \leq a < v$ and $0 \leq b < v$ define

$$F_{a,b} = \{(x,y) \in G_a(V) \times G_p(V) \,|\, E(x) \subseteq E(y)\} \text{ if } a \leqq b$$

$$F_{a,b} = \{(x,y) \in G_a(V) \times G_b(V) \,|\, E(x) \supseteq E(y)\} \text{ if } a > b.$$

Let $\pi: F_{a,b} \to G_a(V)$ and $\tau: F_{a,b} \to G_b(V)$ be the natural projection. For $y \in G_b(V)$, define $S_y = \tau\pi^{-1}(y)$.

Proposition 2.1. $\mathfrak{A}_{a,b}(V) = \{S_y\}_{y \in G_b(V)}$ is an admissible family on $G_a(V)$.

Proof: Obviously, $F_{a,b}$ is closed and locally given by holomorphic equations. Therefore, $F_{a,b}$ is a compact, analytic subset of $G_a(V) \times G_b(V)$. Let $GL(V) = \{\alpha: V \to V | \alpha \text{ linear isomorphism}\}$ be the general linear group of V. Then GL(V) acts on $G_a(V)$ by $\alpha(E(x)) = E(\alpha(x))$. If $(x,y) \in F_{a,b}$ and $\alpha \in GL(V)$, then $(\alpha(x),\alpha(y)) \in F_{a,b}$. Then, GL(V) acts as a transitive group of biholomorphic maps on $F_{a,b}$. Since $F_{a,b}$ is smooth somewhere, it is smooth everywhere. Hence, $F_{a,b}$ is a smooth complex submanifold of $G_a(V) \times G_b(V)$. Obviously, the projections π and τ are surjective, proper, holomorphic and commute with the action of GL(V). By Sards theorem π and τ are regular at least along one of its fibers; hence by the action of GL(V), they are regular everywhere; q.e.d.

3. Example.

Choose $a = 0$ and $b = p$ and $v = n$ in example 2. Then $G_a(V) = \mathbb{P}(V)$ and $S_y = \ddot{E}(y)$. The family $\mathfrak{A}_p(V) = \mathfrak{A}_{0,p}(V)$ is the family of p-dimensional projective linear subspaces in P(V). This case was extensively treated in [28] and [30] and is the foundation for the more general theory presented here.

<u>4. Example.</u>

Choose b = 0 in example 3. Then $G_p(V) = \mathbb{P}(V)$ and S_y is the set of all a-dimensional projective linear subspaces in V containing y. This case was treated by Bott and Chern [1] using the language of holomorphic vector bundles.

Let M and N be complex manifolds with dim M = m and dim N = n. Let f: M → N be an holomorphic map. Let $\mathfrak{A}$ be an admissible family on N of codimension s = n - p, given by the defining triplet $N \xleftarrow{\tau} F \xrightarrow{\pi} A$. Define

$$f^*(F) = \{(z,y) \in M \times F \,|\, f(z) = \tau(y)\}$$

Obviously, $f^*(F)$ is an analytic subset of M x F. The natural projections $\tilde{f}: f^*(F) \to F$ and $\sigma: f^*(F) \to M$ as well as $\hat{f} = \tilde{f} \circ \pi$ are holomorphic with

$$\tau \circ \tilde{f}(z,y) = \tau(y) = f(z) = f \circ \sigma(z,y).$$

Hence, the <u>fundamental diagram</u>

$$\begin{array}{llcll} \hat{f}: & f^*(F) & \xrightarrow{\tilde{f}} & F & \xrightarrow{\pi} A \\ & \downarrow \sigma & \circ & \downarrow \tau & \\ & M & \xrightarrow{f} & N & \end{array}$$

is commutative. Let t be the fiber dimension of τ.

Proposition 2.2. The analytic set $f^*(F)$ is a smooth complex submanifold of dimension $m + t$ of $M \times F$. The holomorphic map $\sigma: f^*(F) \to M$ is proper, surjective and regular. The fiber dimension of σ is t. The restriction $\tilde{f}: \sigma^{-1}(z) \to \tau^{-1}(f(z))$ is biholomorphic for every $z \in M$. For every $a \in A$, the restriction

$$\sigma_a = \sigma | \hat{f}^{-1}(a): \hat{f}^{-1}(a) \longrightarrow f^{-1}(S_a)$$

is bijective and holomorphic.

Proof. Take $(z_0, w_0) \in f^*(F) \subseteq M \times F$. Define $x_0 = \tau(w_0)$ and $y_0 = \pi(z_0)$. Then the following commutative diagram exists:

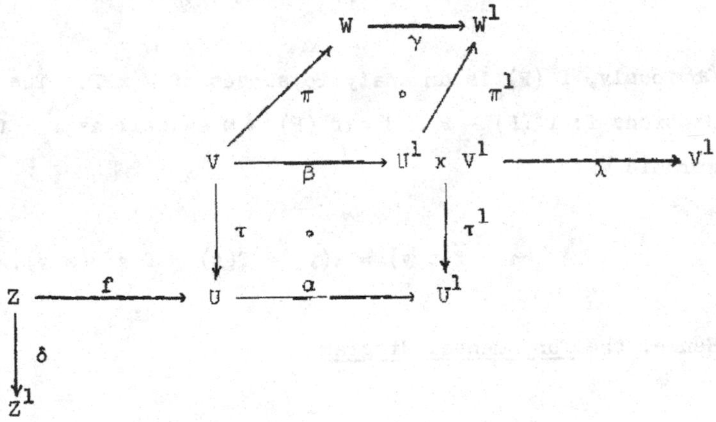

Here, Z, U, U^1, V^1, V, W, W^1 are open with $z_0 \in Z \subseteq M$, and $w_0 \in V \subseteq F$, and $y_0 \in W = \pi(V) \subseteq A$, and $x_0 \in U = \tau(V) \subseteq N$, and

and $U^1 \subseteq \mathbb{C}^n$, and $V^1 \subseteq \mathbb{C}^t$, and $W^1 \subseteq \mathbb{C}^k$, and $Z^1 \subseteq \mathbb{C}^m$. The maps α, β, γ and δ are biholomorphic. The maps τ and λ are the natural projections. Moreover, $\pi^1 = \gamma \circ \pi \circ \beta^{-1}$ is regular and surjective. Then

$$F^1 = \{(\mathfrak{z}, \alpha \circ f \circ \delta^{-1}(\mathfrak{z}), \mathfrak{y}) \mid (\mathfrak{z}, \mathfrak{y}) \in Z^1 \times V^1\}$$

is a smooth, $(m + t)$-dimensional, complex submanifold of $Z^1 \times U^1 \times V^1$.

Obviously, $\delta \times \beta\colon Z \times V \to Z^1 \times U^1 \times V^1$ is biholomorphic. If $(z,w) \in f^*(F) \cap (Z \times V)$, then

$$\begin{aligned}(\delta \times \beta)(z,w) &= (\delta(z), \tau^1 \circ \beta(w), \lambda \circ \beta(w)) \\ &= (\delta(z), \alpha \circ \tau(w), \lambda \circ \beta(w)) \\ &= (\delta(z), \alpha \circ f(z), \lambda \circ \beta(w)) \in F^1\end{aligned}$$

If $u = (\mathfrak{z}, \alpha \circ f \circ \delta^{-1}(\mathfrak{z}), \mathfrak{y}) \in F^1$, define $z = \delta^{-1}(\mathfrak{z}) \in Z$ and $w = \beta^{-1}(\alpha \circ f \circ \delta^{-1}(\mathfrak{z}), \mathfrak{y}) \in V$. Then

$$\alpha(f(z)) = \tau^1(\beta(w)) = \alpha(\tau(w))$$

which implies $f(z) = \tau(w)$. Hence $(z,w) \in f^*(F) \cap (Z \times V)$. Moreover,

$$(\delta \times \beta)(z,w) = (\mathfrak{z}, \alpha \circ f(z), \mathfrak{y}) = u$$

Hence, $\delta \times \beta\colon f^*(V) \cap (Z \times V) \to F^1$ bijectively. Therefore, $f^*(F) \cap (Z \times V)$ is a smooth submanifold of dimension $m + t$ of $Z \times V$.

Let $\mu: F^1 \to Z^1 \times V^1$ and $\sigma^1: Z^1 \times V^1 \to Z^1$ be the natural projections. Obviously, μ is biholomorphic and σ^1 is regular. Because δ is biholomorphic and because $\delta \circ \sigma = \sigma^1 \circ \mu \circ (\delta \times \beta)$, the map $\sigma: f^*(V) \cap (Z \times V) \to Z$ is regular.

Take $z \in M$, then $y \in F$ exists such that $f(z) = \tau(y)$. Hence $(z,y) \in f^*(F)$ and $z = \sigma(z,y)$. Therefore, σ is surjective. If K is a compact subset of M, then $\sigma^{-1}(K)$ is a closed subset of the compact set $K \times \tau^{-1}(f(K))$. Hence, σ is proper.

Obviously, $\tilde{f}: \sigma^{-1}(z) \to \tau^{-1}(f(z))$ is biholomorphic. Take $a \in A$. If $(z,y) \in \hat{f}^{-1}(a)$, then $f(z) = \tau(y)$ and $\pi(y) = a$. Hence $f(z) \in S_a$, which implies $\sigma(z,y) = z \in f^{-1}(S_a)$. If $z \in f^{-1}(S_a)$, then $y \in \pi^{-1}(a)$ with $f(z) = \tau(y)$ exists. Hence $(z,y) \in \hat{f}^{-1}(a)$. If $(z,y) \in \hat{f}^{-1}(a)$ and $(z,y^1) \in \hat{f}^{-1}(a)$, then $f(z) = \tau(y) = \tau^1(y^1)$ with $y \in \pi^{-1}(a)$ and $y^1 \in \pi^{-1}(a)$. Because $\tau: \pi^{-1}(a) \to S_a$ is injective, $y = y^1$. Therefore $\sigma: \hat{f}^{-1}(a) \to f^{-1}(S_a)$ is bijective, q.e.d.

The map f is said to be <u>adapted to $\mathfrak{A}$ at z_0</u> $\in M$ for $a \in A$, if and only if open neighborhoods U of z_0 in M and V of a in A exist, such that $\dim_z f^{-1}(S_y) = q = m - s$ for all $z \in f^{-1}(S_y) \cap V$ and for all $y \in A$. Obviously, if $z_0 \notin f^{-1}(S_a)$, then f is adapted to $\mathfrak{A}$ at z_0 for a.

<u>Proposition 2.3.</u> <u>Let $z_0 \in f^{-1}(S_a)$, then f is adapted to $\mathfrak{A}$ at z_0 for a, if and only if $\sigma_a^{-1}(z_0)$ is contained in an open subset W of $f^*(F)$ such that $\hat{f}|W$ is open. (Observe that $(z_0,y_0) = \sigma_a^{-1}(z_0)$ is the only point of $f^*(F)$ in $\hat{f}^{-1}(a)$ which is mapped by σ onto z_0.)</u>

<u>Proof</u>. a) Suppose that f is adapted to $\mathfrak{A}$ at z_0 for a, and that $f^{-1}(S_y) \cap U$ is empty or pure q-dimensional for every $y \in V$, where U and V are open neighborhoods of z_0, respectively a. Then $\sigma_y: \hat{f}^{-1}(y) \cap \sigma^{-1}(U) \to f^{-1}(S_y) \cap U$ is holomorphic and bijective. Hence $\hat{f}^{-1}(y) \cap \sigma^{-1}(U)$ is empty or pure q-dimensional for every $y \in V$. According to Remmert [18], the map $\hat{f}$ is open on the neighborhood $\sigma^{-1}(U) \cap \hat{f}^{-1}(V)$ of $\sigma_a{}^{-1}(z_0)$.

b) Suppose, that $\hat{f}|W$ is open for some open neighborhood W of $(z_0,y_0) = \sigma_a{}^{-1}(z_0)$. Open neighborhoods U of z_0 and V of a exist such that

$$\tilde{W} = f^*(F) \cap (U \times \pi^{-1}(V)) \subseteq W.$$

For, if this would be wrong, a sequence $\{(x_\nu,w_\nu)\}_{\nu \in \mathbb{N}}$ of points in $f^*(F) - W$ would exist, such that $x_\nu \to z_0$ for $\nu \to \infty$ and $\pi(w_\nu) \to a$ for $\nu \to \infty$. Let K be a compact neighborhood of z_0, then $x_\nu \in K$ for $\nu \geqq \nu_1$ and $(x_\nu,w_\nu) \in \sigma^{-1}(K)$ for $\nu \geqq \nu_1$, where $\sigma^{-1}(K)$ is compact. Hence, it can be assumed that $w_\nu \to w$ for $\nu \to \infty$. Because $f^*(F)$ is closed in $M \times F$, also $(z_0,w) \in f^*(F)$. Moreover $\hat{f}(z_\nu,w_\nu) = \pi(w_\nu)$ converges to $\hat{f}(z_0,w) = \pi(w) = a$. Hence $w = y_0$ and $(z_0,w) \in W$. Therefore $(z_\nu,w_\nu) \in W$ for some ν, which is wrong. Therefore, U, V and $\tilde{W}$ exist.

If $z \in f^{-1}(S_y) \cap U$ with $y \in V$, then $f(z) = \tau(w)$ for some $w \in \pi^{-1}(y)$. Hence $(z,w) \in \tilde{W}$ and $\sigma(z,w) = z$. Therefore $\sigma_y: \hat{f}^{-1}(y) \cap \tilde{W} \to f^{-1}(S_y) \cap U$ is bijective. According to Remmert [18],

$f^{-1}(y) \cap \tilde{W}$ is either empty or has the pure dimension

$$\dim f^*(F) - \dim A = (m + t) - (n + t - p) = m - s = q.$$

Hence $f^{-1}(S_y) \cap U$ is either empty or pure q-dimensional if $y \in V$; i.e., f is adapted to $\mathfrak{A}$ at z_0 for a; q.e.d.

For any subset K of M define

$$L(K) = \{a \in A \mid f \text{ adapted to } \mathfrak{A} \text{ at all } x \in K \text{ for } a\}.$$

If $K_1 \subseteq K_2$, then $L(K_2) \subseteq L(K_1)$. If $K = \bigcup_{\lambda\in\Lambda} K_\lambda$ then $L(K) = \bigcap_{\lambda\in\Lambda} L(K_\lambda)$.

Proposition 2.4. For every $K \subseteq M$, the set $A - L(K)$ has measure zero. If K is compact, then $A - L(K)$ is compact and $L(K)$ open.

Proof. Let E be the set of points $z \in f^*(F)$ such that $\hat{f}|U$ is not open for any neighborhood U of z. According to Remmert [18], the set E is analytic. Proposition 2.3 implies that $a \in A - L(K)$ if and only if $(x,y) \in E \cap \hat{f}^{-1}(a)$ exists such that $x \in K$. Hence

$$A - L(K) = \hat{f}(E \cap \sigma^{-1}(K))$$

If K is compact, then $\sigma^{-1}(K)$ is compact and E closed. Hence $A - L(K)$ is compact and $L(K)$ is open. Let S be the set of points in $f^*(F)$ where $\hat{f}$ is not regular. By Sards Theorem $\hat{f}(S)$ is a set of measure zero. Obviously, $S \supseteq E$. Hence $A - L(M) = \hat{f}(E)$ is a set of measure zero. Trivially, $A - L(K) \subseteq A - L(M)$ is also a set of measure zero.
q.e.d.

Because σ is proper, $\sigma(E)$ is an analytic subset of M. Hence $M - \sigma(E) = \{x \mid f$ adapted to $\mathfrak{A}$ at x for all $a \in A\}$ is the complement of an analytic set. However, $M = \sigma(E)$ may be possible.

If f is adapted to $\mathfrak{A}$ at z for $a \in A$, an <u>intersection multiplicity</u> $\nu_f^a(z)$ shall be defined. If $f(z) \notin S_a$, define $\nu_f^a(z) = 0$. If $f(z) \in S_a$, one and only one point $y \in \pi^{-1}(a)$ exists such that $(z,y) \in f^*(F)$ and $\hat{f}$ is open in a neighborhood of this point. Hence, the multiplicity $\nu_{\hat{f}}(z,y)$ of f at (z,y) is defined (See [26]). Set $\nu_f^a(z) = \nu_{\hat{f}}(z,y)$.

The holomorphic map $f\colon M \to N$ is said to be <u>almost adapted to $\mathfrak{A}$</u> if and only if for each component M_λ of M a point $x_\lambda \in M_\lambda$ and for each component T_μ of $\tau^{-1}(f(x_\lambda))$ a point $a_{\lambda\mu} \in \pi(T_\mu) \subseteq A$ exists[6] such that f is adapted to $\mathfrak{A}$ at x_λ for each $a_{\lambda\mu}$. Observe that $\tau^{-1}(y)$ has only finitely many components for $y \in N$. Now, $\tau\colon F \to N$ is a differentiable fiber bundle. If N is connected, all fibers of τ are connected if and only if one fiber is connected. Hence the following Lemma is true.

<u>Lemma 2.5.</u> <u>If M is connected, and if at least one fiber of $\tau\colon F \to N$ is connected, then f is almost adapted to $\mathfrak{A}$ if and only if a point $x_0 \in M$ and a point $a_0 \in A$ with $f(x_0) \in S_{a_0}$ exists such that f is adapted to $\mathfrak{A}$ at x_0 for a_0.</u>

<u>Proposition 2.6.</u> <u>Let $f\colon M \to N$ be almost adapted to $\mathfrak{A}$. Let S be the set of points of $f^*(F)$ where $\hat{f}$ is not regular. Let Ξ be the set of points $z \in f^*(F)$ such that $\hat{f}|U$ is not open for each open</u>

neighborhood U of z. Then E and S are thin analytic subsets of $f^*(F)$.

Proof. Let H be a component of $f^*(F)$. Because $\sigma: f^*(F) \to M$ is surjective, proper, holomorphic and regular, also $\sigma: H \to M$ is proper, holomorphic and regular. Then $\sigma(H)$ is a component of M. Pick $x \in \sigma(H)$ such that for each component T_μ of $\tau^{-1}(f(x))$ a point $y_\mu \in T_\mu$ exists such that f is adapted to $\mathfrak{A}$ at x for $a_\mu = \pi(y_\mu)$. Then $f(x) = \tau(y_\mu)$, hence $(x,y_\mu) \in f^*(F)$. The map $\tilde{f}_x: \sigma^{-1}(x) \to \tau^{-1}(f(x))$ is biholomorphic. Hence $\tilde{f}_x^{-1}(T_\mu) = T_\mu^1$ are the components of $\sigma^{-1}(x)$. For some μ, the component T_μ^1 is contained in H and $(x,y_\mu) \in H$. Hence $\hat{f}$ is open in a neighborhood of (x,y_μ). Hence $\hat{f}(H)$ contains a_μ as an interior point. Because $\hat{f}(S)$ is a set of measure zero, S does not contain H. Hence S is a thin analytic subset of $f^*(F)$. Because E is analytic and $E \subseteq S$, this is also true for E. q.e.d.

Let C be the set of all $x \in M$ with the property: "A point $a \in A$ exists (depending on x) such that $f(x) \in S_a$ and such that f is adapted to $\mathfrak{A}$ at x for a".

Proposition 2.7. If $f: M \to N$ is almost adapted to $\mathfrak{A}$, then C is dense in M.

Proof. Let U be any open, non-empty subset of M. Then $\sigma^{-1}(U) \neq \emptyset$ is open in $f^*(F)$. Hence, a point (x,y) in $\sigma^{-1}(U) - E$ exists according to Proposition 2.6. Then $\sigma(x,y) = x \in U$. Set $\hat{f}(x,y) = \pi(y) = a \in A$. Because $\hat{f}$ is open in a neighborhood of (x,y), the map f is adapted to $\mathfrak{A}$ at x for a by Proposition 2.3. Moreover $f(x) = \tau(y)$ belongs to $\tau\pi^{-1}(a) = S_a$. Therefore $x \in U \cap C$, q.e.d.

§3. The definition of the proximity form

At first several concepts and notations have to be introduced:

a) Non-negative forms. Let M be a complex manifold of dimension m. Let χ be a form of bidegree (q,q) on M with $0 < q \leqq m$. For $x \in M$, let $\mathcal{L}_x(q)$ be the set of all smooth, q-dimensional, complex submanifolds L of M with $x \in L$. Let $j_L: L \to M$ be the inclusion map. Then $j_L^*(\chi)$ is a form of top degree on L. The form χ is said to be non-negative (positive) at x, if $j_L^*(\chi)(x) \geqq 0$ (resp. > 0) for all $L \in \mathcal{L}_x(q)$. The form χ is said to be non-negative (positive) if and only if it is non-negative (resp. positive) at every $x \in M$.

If the forms χ_1 and χ_2 of bidegree (q,q) and the functions f_1, f_2 are non-negative at $x \in M$, so is $f_1\chi_1 + f_2\chi_2$. If the form χ_1 of bidegree (q,q) and the form χ_2 of bidegree (1,1) on M are non-negative (respectively positive) at x, so is $\chi_1 \wedge \chi_2$. If φ is a form of bidegree (q,0) on M, then $(i)^{q^2} \varphi \wedge \overline{\varphi}$ is non-negative. A form χ of bidegree (q,q) is non-negative at x, if and only if for any collection $\alpha_{q+1}, \ldots, \alpha_m$ of forms of bidegree (1,0), the form $(i)^{(m-q)^2} \chi \wedge \alpha_{q+1} \wedge \overline{\alpha}_{q+1} \wedge \cdots \wedge \alpha_m \wedge \overline{\alpha}_m$ is non-negative at x.[7]

If φ is a continuous real form of bidegree (q,q) on M, if ψ is a continuous, positive form of bidegree (q,q) on M, and if K is a compact subset of M, a constant $c > 0$ exists such that $\varphi + c\psi$ is positive.[8]

A form χ of bidegree (q,q) is said to be strictly non-negative at $x \in M$, if there exist non-negative forms $\varphi_{\mu\nu}$ of bidegree (1,1) such that

$$\chi = \sum_{\mu=1}^{r} \varphi_{\mu 1} \wedge \cdots \wedge \varphi_{\mu q} \qquad \text{at } x.$$

If this is the case, and if the form ψ of bidegree (p,p) is non-negative at x, then $\chi \wedge \psi$ is non-negative at x.

Let V be a complex vector space of dimension n. Let $(\,|\,) : V \times V \to V$ be an <u>hermitian product</u> on V, i.e., $(\cdot|y)$ is linear over C, and $(x|y) = \overline{(y|x)}$ and $(x|x) > 0$ if $x \neq 0$. Define $|x| = \sqrt{(x|x)}$. Then the form υ and υ_p defined by

$$\upsilon(x) = \frac{1}{4} d^{\perp} d |x|^2$$

$$\upsilon_p = \frac{1}{p!} \upsilon \wedge \cdots \wedge \upsilon \qquad \text{(p-times)}$$

are positive on V. The forms ω and ω_p defined by

$$\omega(x) = \frac{1}{4} d^{\perp} d \log |x|^2$$

$$\omega_p = \frac{1}{p!} \omega \wedge \cdots \wedge \omega \qquad \text{(p-times)}$$

are non-negative on V - {0}. One and only one form $\ddot{\omega}$ of bidegree (1,1) exists on $\mathbb{P}(V)$ such that $\mathbb{P}^*(\ddot{\omega}) = \omega$ on V - {0}. This form $\ddot{\omega}$ is is positive on $\mathbb{P}(V)$ and the associated exterior form of a Kaehler metric on $\mathbb{P}(V)$. Define $\ddot{\omega}_p = \frac{1}{p!} \ddot{\omega} \wedge \cdots \wedge \ddot{\omega}$ (p-times). Then $\mathbb{P}^*(\ddot{\omega}_p) = \omega_p$ on V-{0}.

On $\mathbb{C}^n$, the hermitian product

$$(x|y) = \sum_{\mu=1}^{n} x_\mu \overline{y}_\mu$$

will always be used, where $x = (x_1,\ldots,x_n)$ and $y = (y_1,\ldots,y_n)$.

b). <u>Some abbreviations</u>. For each positive integer p define

$$\Delta_p = \{x \in \mathbb{N} \mid 1 \leq x \leq p\}$$

If p and q are positive integers with $p \leq q$, define

$$T(p,q) = \{\mu\colon \Delta_p \to \Delta_q \mid \mu \text{ increasing and injective}\}$$

If $e_1,\ldots,e_q$ are elements of a vector space V, and if $\mu \in T(p,q)$, define

$$e_\mu = e_{\mu(1)} \wedge \cdots \wedge e_{\mu(p)}$$

If $\mu \in T(p,q)$ and $\nu \in T(p,q)$ and if V has a conjugation, define

$$e_{\mu\nu} = \frac{(i)^{p^2}}{2^p}\, e_\mu \wedge \overline{e}_\nu = \overline{e}_{\nu\mu}$$

$$= \left(\frac{i}{2}\right)^p e_{\mu(1)} \wedge \overline{e}_{\nu(1)} \wedge \cdots \wedge e_{\mu(p)} \wedge \overline{e}_{\nu(p)}.$$

This notation extends to vector bundles and their sections.

c). <u>Boundary manifolds</u>. If not otherwise specified, "<u>differentiable</u>", "<u>diffeomorphism</u>", etc., are meant to be of class C^∞. A manifold is assumed to be oriented, para-compact and pure dimensional. A diffeomorphism is assumed to be orientation preserving.

Let M be a differentiable manifold of dimension m. Let H be open in M. A differentiable manifold S of dimension m-1 and contained in $\overline{H} - H$ is said to be a <u>boundary manifold of</u> H if and only if for every $a \in S$ an open neighborhood U of a and open neighborhoods $U' \subseteq \mathbb{R}$ and $U'' \subseteq \mathbb{R}^{m-1}$ of 0 and an (orientation preserving) diffeomorphism $\alpha: U \to U' \times U''$ with $\alpha(a) = 0$ exist such that

1. $\alpha = (x_1,\ldots,x_m)$ on U.
2. $\beta = (x_2,\ldots,x_m) : U \cap S \to U''$ is an (orientation preserving) diffeomorphism.
3. $U \cap H = \{z \in U | x_1(z) < 0\}$ and $U \cap S = \{z \in U | x_1(z) = 0\}$.

e) <u>Sets of measure zero on analytic subsets</u>. Let A be an analytic subset of the complex manifold M of pure dimension q. Let Let A' be the set of simple points of A. Let S be a subset of A. If $q > 0$, then S is said to have <u>measure zero</u> if and only if $A' \cap S$ is a set of measure zero on the complex manifold A'. If $q = 0$, then S is said to have measure zero, if and only if S is empty. (See [21] and Lelong [11]).

f) <u>Support on subsets</u>. Let ω be a form of degree p on the differentiable or complex manifold M. Let $S \subseteq M$. Then $\operatorname{supp} \omega | S$ is the closure in S of the set $\{x \in S | \omega(x) \neq 0\}$. Observe, that $S \cap \operatorname{supp} \omega \neq \operatorname{supp} \omega | S$ in general. If S is a submanifold of M and if $j: S \to M$ is the inclusion, then $\operatorname{supp} \omega | S \neq \operatorname{supp} j^*(\omega)$ is possible

g) <u>Lipschitz function</u>. Let M be a differentiable manifold.

A function $\psi: M \to R$ satisfies a (local) Lipschitz condition, if and only if for every $a \in M$ a diffeomorphism $\alpha: U \to U'$ of an open neighborhood U of a onto an open subset U' of $\mathbb{R}^m$ exists such that $\psi \circ \alpha^{-1}$ satisfies a Lipschitz condition on U'. Then $d\psi$ exists almost everywhere, which is also true for $d^{\perp}\psi$ on complex manifolds.

Lemma 3.1. Let M be a complex manifold of dimension m. Let p, q, r be non-negative integers with $p + q + r = m - 1$. Let ψ, χ, λ forms of bidegree (p,p), (q,q), (r,r) respectively. Then

$$d\psi \wedge d^{\perp}\lambda \wedge \chi = d\lambda \wedge d^{\perp}\psi \wedge \chi$$

Proof. The consideration of bidegrees implies

$$\begin{aligned} d\psi \wedge d^{\perp}\lambda \wedge \chi &= i(\partial\psi + \bar{\partial}\psi) \wedge (\partial\lambda - \bar{\partial}\lambda) \wedge \chi \\ &= i\partial\psi \wedge \partial\lambda \wedge \chi - i\bar{\partial}\psi \wedge \bar{\partial}\lambda \wedge \chi \\ &\quad - i\partial\psi \wedge \bar{\partial}\lambda \wedge \chi - i\partial\lambda \wedge \bar{\partial}\psi \wedge \chi \\ &= -i\partial\psi \wedge \bar{\partial}\lambda \wedge \chi - i\partial\lambda \wedge \bar{\partial}\psi \wedge \chi \\ &= d\lambda \wedge d^{\perp}\psi \wedge \chi \end{aligned}$$

q.e.d.

Let N be a complex manifold of dimension n. Let $\mathfrak{a}$ be an admissible family of codimension s on N given by the triplet $N \xleftarrow{\tau} F \xrightarrow{\pi} A$. Let Ω be a form of class C^{∞} and of bidegree (s,s) on N. Take $a \in A$. A non-negative form Λ_a of class C^{∞} and of bidegree (s-1,s-1) on $N-S_a$ is said to be a <u>proximity form of Ω for a</u> if and only if

$$d^{\perp}d\Lambda_a = \Omega$$

on $N - S_a$ and if in addition the "Residue Theorem" and the "Singular Stokes Theorem" hold.

Both of these theorems shall be stated now: Let M be any complex manifold of dimension $m \geqq s$. Define $q = m - s$. Let $f: M \to N$ be a holomorphic map. Let χ be a form of class C^{∞} and of bidegree (q,q) on M. Let H be a non-empty, open relative compact subset of M, whose boundary S is either empty or a boundary manifold of H. Let $j: S \to M$ be the inclusion. Suppose that f is adapted to $\mathfrak{a}$ for a at every point $x \in f^{-1}(S_a) \cap \bar{H}$. Let $\psi: M \to \mathbb{R}$ be a continuous function and let $T = \operatorname{supp}(\psi\chi)|S$.

<u>Residue Theorem.</u>[9] <u>Suppose that ψ satisfies a Lipschitz condition. Suppose that $T \cap f^{-1}(S_a)$ is a set of measure zero on $f^{-1}(S_a)$. Assume that $j^*(\psi d^{\perp}f^*(\Lambda_a) \wedge \chi)$ is integrable over S. Then all the following integrals exist and satisfy the identity</u>

$$\int_S \psi d^{\perp}f^*(\Lambda_a) \wedge \chi + \int_H \psi f^*(\Omega) \wedge \chi =$$

$$\int_{H \cap f^{-1}(S_a)} \nu_f^a \psi\chi + \int_H d\psi \wedge d^{\perp}f^*(\Lambda_a) \wedge \chi$$

$$+ \int_H \psi \wedge d^{\perp}f^*(\Lambda_a) \wedge d\chi$$

(If $q = 0$, then the integral over $H \cap f^{-1}(S_a)$ is a sum over this set).

<u>Singular Stokes Theorem.</u>[10] <u>Suppose that $\psi|\overline{H}$ is of class C^2. Suppose that either a) or b) hold</u>

a) The form $j^*(f^*(\Lambda_a) \wedge d^{\perp}\psi \wedge \chi)$ is integrable over S.

b) On each connectivity component of S, the form $j^*(f^*(\Lambda_a) \wedge d^{\perp}\psi \wedge \chi)$ is either non-negative or non-positive.

Then all the following integrals exist and satisfy the identity

$$\int_H df^*(\Lambda_a) \wedge d^{\perp}\psi \wedge \chi$$

$$= \int_S f^*(\Lambda_a) \wedge d^{\perp}\psi \wedge \chi - \int_H f^*(\Lambda_a) \wedge dd^{\perp}\psi \wedge \chi$$

$$+ \int_H f^*(\Lambda_a) \wedge d^{\perp}\psi \wedge d\chi$$

(Among others, this means that b) implies a)). (Observe that $d\psi$ and $d^{\perp}\psi$ are defined on $\overline{H}$.)

The advantage of this axiomatic definition is that it will lead quickly to the First Main Theorem and that the reader is spared a considerable construction job for some short while.

The disadvantage is, that the definition is logically rather complicated. Moreover, it will be impossible to construct a proximity form in this general situation.

A proximity form will be constructed for the point family $\mathfrak{A}_N$ if $N = A$ is a compact Kaehler manifold and if Ω is the volume element of the Kaehler metric normalized such that $\int_N \Omega = 1$. If $\mathfrak{A}$ is an admissible family given by the triplet $N \xleftarrow{\tau} F \xrightarrow{\pi} A$, if A is a compact Kaehler manifold with normalized volume element, then it will be possible to construct a form Λ_a such that the First Main Theorem still holds, but where Λ_a is only a <u>"weak" proximity form</u> this means that $d^{\perp}d\Lambda_a = \Omega$ on $N - S_a$, and that the "Residue Theorem" and the "Singular Stokes Theorem" hold under the following additional assumptions:

1) In the Residue Theorem, $\psi | S = 0$.

2) In the Singular Stoke's Theorem, χ is strictly non-negative. ψ is constant on every connectivity component of S, where the constant is a relative maximum or minimum of $\psi | \overline{H}$.

If $\mathfrak{A}_p(V)$ (Example 3, §2), the Levine form is a proximity form. See Levine [14], Chern [2], and [28], [30]. If A is a compact homogeneous Kaehler manifold a proximity form was constructed by Hirschfelder [7]. If q = 0, a proximity form was constructed by Wu [33], for the point family. It is also a proximity form for q > 0 for the point family, as Hirschfelder [7a] has shown.

Several remarks about the Residue Theorem and the Singular Stoke's Theorem shall be made:

<u>1. Remark</u>. The so called "<u>Unintograted First Main Theorem</u>" is obtained from the Residue Theorem by setting $\psi = 1$ and assuming $d\chi = 0$:

$$\int_H f^*(\Omega) \wedge \chi = \int_{H \cap f^{-1}(S_a)} \nu_f^a \chi - \int_S d^\perp f^*(\Lambda_a) \wedge \chi .$$

Even if, $\chi \geqq 0$ and $\Omega \geqq 0$ is assumed, the boundary integral has an unknown sign.

2. Remark. Suppose that $\psi = 1$, $d\chi = 0$ and that $H = M$ is compact. Then

$$\int_M f^*(\Omega) \wedge \chi = \int_{f^{-1}(S_a)} \nu_f^a \chi$$

3. Remark. If $\psi = \chi = 1$, if $q = 0$ and $s = n$ then

$$\int_N d^\perp d\Lambda_a = \int_N \Omega = 1$$

4. Remark. Lemma 3.1 implies

$$df^*(\Lambda_a) \wedge d^\perp \psi \wedge \chi = d\psi \wedge d^\perp f^*(\Lambda_a) \wedge \chi$$

Lemma 3.2.[11] Let H be an open subset of the m-dimensional complex manifold M. Let S be a boundary manifold of H. Let $z_0 \in S$. Let $\psi: M \to \mathbb{R}$ be a function of class C^1 on M. Suppose that an open neighborhood U of z_0 in M exists such that $\psi(z) \geqq \psi(z_0)$ for $z \in U \cap \bar{H}$. Let χ be a non-negative form of bidegree $(m-1,m-1)$ on M. Let $j: S \to M$ be the inclusion map. Then

$$j^*(d^\perp \psi \wedge \chi)(z_0) \geqq 0.$$

Proof. Without loss of generality, it can be assumed that M is an open neighborhood of $z_0 = 0 \in \mathbb{C}^m$ and that $\psi(z_0) = 0$. Let $z_1,\ldots,z_m$ be the coordinates of $\mathbb{C}^m$ and let $x_1,\ldots,x_{2m}$ be the real coordinates of $\mathbb{C}^m$ with $z_\mu = x_{2\mu-1} + ix_{2\mu}$ for $\mu = 1,\ldots,m$.

Moreover, it can be assumed, that $U = U' \times U''$ where U' is an open neighborhood of $0 \in \mathbb{R}$ and U'' is an open neighborhood of $0 \in \mathbb{R}^{2m-1}$ and where a C^∞ function f on U'' exists such that $f(0) = 0$ and $(df)(0) = 0$ and

$$S \cap U = \{(f(t),t) \mid t \in U''\}$$

$$H \cap U = \{(x_1,t) \in U' \times U'' \mid x_1 < f(t)\}.$$

For $1 \leqq \mu \leqq m$ and $1 \leqq \nu \leqq m$ with $\mu \neq \nu$ define

$$\zeta_{\mu\nu} = \prod_{\substack{\lambda=1 \\ \mu\neq\lambda\neq\nu}}^{m} dz_\lambda \wedge d\bar{z}_\lambda$$

$$\zeta_\mu = \prod_{\substack{\lambda=1 \\ \lambda\neq\mu}}^{m} dz_\lambda \wedge d\bar{z}_\lambda$$

Then

$$\chi = \left(\tfrac{i}{2}\right)^{m-1} \sum_{\substack{\mu,\nu=1 \\ \mu\neq\nu}}^{m} a_{\mu\nu} d\bar{z}_\mu \wedge dz_\nu \wedge \zeta_{\mu\nu}$$

$$+ \left(\tfrac{i}{2}\right)^{m-1} \sum_{\mu=1}^{m} a_{\mu\nu}\zeta_\mu$$

Because χ is non-negative, $a_{\mu\nu} = \overline{a}_{\nu\mu}$ and

$$\sum_{\mu=1}^{m} a_{\mu\nu} h_\mu \overline{h}_\nu \geqq 0$$

if $h_1,\ldots,h_m$ are complex numbers. Especially $a_{\mu\mu} \geqq 0$ for $\mu = 1,\ldots,n$.

On U'' define g by $g(t) = \psi(f(t),t)$ for $t \in U''$. Then $g(0) = 0$ and $g(t) \geqq 0$ for $t \in U''$. Hence $(dg)(0) = 0$, which means

$$0 = \psi_{x_1}(0) \cdot f_{x_\nu}(0) + \psi_{x_\nu}(0) = \psi_{x_\nu}(0)$$

for $\nu = 2,\ldots,2m$. Hence $\psi_{z_\mu}(0) = \psi_{\overline{z}_\mu}(0) = \Theta$ for $\mu = 2,\ldots,m$ and $\psi_{z_1}(0) = \frac{1}{2}\psi_{x_1}(0) = \psi_{\overline{z}_1}(0)$. Therefore

$$(j^* d^\perp \psi)(0) = \frac{1}{2}\psi_{x_1}(0) j^*(dz_1 - d\overline{z}_1)$$

Because $df(0) = 0$,

$$j^*(dz_\mu)(0) = (dz_\mu)(0) \text{ if } 1 < \mu \leqq m$$

$$j^*(dz_1)(0) = i(dx_2)(0) = -j^*(d\overline{z}_1)(0)$$

Hence

$$j^*(d^{\perp}\psi \wedge \chi)(0) = (\tfrac{i}{2})^m \psi_{x_1}(0) j^*((dz_1 - d\bar{z}_1) \wedge \zeta_{11})(0)$$

$$= -\psi_{x_1}(0)(dx_2 \wedge \ldots \wedge dx_{2m})(0).$$

A number $\varepsilon > 0$ exists such that $(x_1, 0, \ldots, 0) \in H \cap U$ if $-\varepsilon < x_1 < f(0) = 0$. Then $\psi(x_1, 0, \ldots, 0) \geqq 0$ for $-\varepsilon < x_1 < 0$. Therefore, $\psi_{x_1}(0) \leqq 0$. Consequently

$$j^*(d^{\perp}\psi \wedge \chi)(0) \geqq 0, \qquad \text{q.e.d.}$$

<u>Lemma 3.3.</u>[12] <u>Let H be an open subset of a differentiable manifold M of dimension m. Let $\psi: M \to \mathbb{C}$ be a continuous function, which assumes on M - H only finitely many values. Suppose that $\psi | \bar{H}$ is of class C^1. Then $\psi: M \to C$ satisfies a Lipschitz condition.</u>

<u>Proof</u>. Without loss of generality, it can be assumed that M is an open subset of $\mathbb{R}^m$. Take $a \in M$. Let B be an open ball with center a such that $\bar{B} \subseteq M$. The ball shall be taken so small that ψ is constant on B - H if B - H $\neq \phi$. Then a constant $c > 0$ exists such that

$$m|\psi_{x_\nu}(x)| \leqq c \qquad \text{if } x \in \bar{B} \cap \bar{H}$$

for $\nu = 1, \ldots, m$. Take any two different points $x \in B$ and $x' \in B$ and define $x(t) = x + t(x' - x)$. Then $x(t) \in B$ for $0 \leqq t \leqq 1$.

<u>1. Case</u>. $x \in B - H$ and $x' \in B - H$. Then

$$|\psi(x) - \psi(x')| = 0 < c|x - x'|$$

<u>2. Case</u>. $x \in H$ and $x' \in B - H$. Define

$$t_0 = \sup \{t | x(u) \in H \quad \text{for} \quad 0 \leqq u \leqq t \leqq 1\}$$

Then $x'' = x(t_0) \in B - H$, because H is open. Hence

$$|\psi(x) - \psi(x')| = |\psi(x) - \psi(x'')|$$

$$= |\sum_{\nu=1}^{m} \psi_{x_\nu}(x(\tau))(x'_\nu - x_\nu)t_0|$$

$$\leqq c|x' - x| .$$

<u>3. Case</u>. $x \in H$ and $x' \in H$. Define t_0 as above. If $t_0 = 1$, the same proof as in 2. applies. If $t_0 < 1$, define

$$t_1 = \inf \{t | x(u) \in H \quad \text{for} \quad 0 \leqq t \leqq u \leqq 1\}.$$

Then $0 < t_0 \leqq t_1 < 1$ and $x'' = x(t_0) \in B - H$ and $x''' = x(t_1) \in B - H$, because H is open. Then

$$\begin{aligned}|\psi(x) - \psi(x')| &= |\psi(x) - \psi(x'') + \psi(x''') - \psi(x')| \\ &\leqq |\psi(x) - \psi(x'')| + |\psi(x''') - \psi(x')| \\ &\leqq c|x - x''| + c|x''' - x'| \\ &\leqq c|x - x'| \qquad \text{q.e.d.}\end{aligned}$$

§4. The First Main Theorem

Let M be a complex manifold of dimension m. Then $B = (G,\Gamma,g,\gamma,\psi)$ is said to be a <u>bump</u>[13)] on M if and only if

1. The subsets G and g of M are open and relative compact with $\phi \neq g \subseteq \overline{g} \subset G \subseteq \overline{G} \subseteq M$.

2. The boundaries Γ of G and γ of g are boundary manifolds of G respectively g.

3. The function $\psi\colon M \to \mathbb{R}$ is continuous on M and constant on $M - G$ and g namely $\psi|(M-G) = 0$ and $\psi|g = R > 0$. For $z \in M$ is $0 \leqq \psi(z) \leqq R$. The restriction $\psi|(\overline{G}-g)$ is of class C^∞.

<u>Theorem 4.1. First Main Theorem.</u> <u>Let M and N be complex manifolds with $m = \dim M$ and $n = \dim N$. Let $\mathfrak{A}$ be an admissible family of codimension s given by a defining triplet $N \xleftarrow{\tau} F \xrightarrow{\pi} A$. Suppose $n - s = p \geqq 0$ and $m - s = q \geqq 0$. Let χ be a strictly non-negative form of class C^∞ and of bidegree (q,q) on M such that $d\chi = 0$. Let $B = (G,\Gamma,g,\gamma,\psi)$ be a bump on M. Let Ω be a non-negative form of class C^∞ and bidegree (q,q) on M. Define the spherical image and the characteristic respectively by</u>

$$A_f(G) = \int_G f^*(\Omega) \wedge \chi$$

$$T_f(G) = \int_G \psi f^*(\Omega) \wedge \chi$$

<u>Both integrals have non-negative integrands.</u>

<u>Take $a \in A$. Suppose, a weak proximity form Λ_a of Ω for $a \in A$</u>

<u>to $\mathfrak{A}$ is given. Suppose that f is adapted to $\mathfrak{A}$ for a at every point of $\overline{G} \cap f^{-1}(S_a)$.</u>

<u>Then the following integrals exist</u>

$$n_f(G,a) = \int_{G \cap f^{-1}(S_a)} v_f^a \chi \geqq 0$$

$$N_f(G,a) = \int_{f^{-1}(S_a)} v_f^a \psi \chi \geqq 0$$

$$m_f(\Gamma,a) = \int_\Gamma f^*(\Lambda_a) \wedge d^\perp \psi \wedge \chi \geqq 0$$

$$m_f(\gamma,a) = \int_\gamma f^*(\Lambda_a) \wedge d^\perp \psi \wedge \chi \geqq 0$$

$$D_f(G,a) = \int_{G-g} f^*(\Lambda_a) \wedge dd^\perp \psi \wedge \chi .$$

<u>The integrands of $n_f(G,a), N_f(G,a), m_f(\Gamma,a)$ and $m_f(\gamma,a)$ are non-negative. Moreover</u>

$$T_f(G) = N_f(G,a) + m_f(\Gamma,a) - m_f(\gamma,a) - D_f(G,a).$$

<u>Names are</u>: $n_f(G,a)$ <u>counting function</u>, $N_f(G,a)$ <u>integrated counting function</u>, $m_f(\Gamma,a)$ <u>proximity function</u>, $m_f(\gamma,a)$ <u>proximity remainder</u>, $D_f(G,a)$ <u>deficit</u>.

<u>Proof</u>. Because $\overline{G}$ is compact, the integrals $T_f(G), n_f(G,a)$ and $N_f(G,a)$ exist. Since $\psi \geqq 0$, since χ is strictly non-negative, since $f^*(\Omega) \geqq 0$, the integrands $v_f^a \chi$ and $v_f^a \psi \chi$ and $\psi f^*(\Omega) \wedge \chi$ are

non-negative. Since $f^*(\Lambda_a) \geqq 0$ and since χ is strictly non-negative, also $f^*(\Lambda_a) \wedge \chi \geqq 0$. Let $j_\Gamma\colon \Gamma \to M$ and $j_\gamma\colon \gamma \to M$ be the inclusions. Lemma 3.2 implies

$$j^*_\Gamma(f^*(\Lambda_a) \wedge d^\perp\psi \wedge \chi) \geqq 0 \qquad \text{on } \Gamma$$

$$j^*_\gamma(f^*(\Lambda_a) \wedge d^\perp\psi \wedge \chi) \geqq 0 \qquad \text{on } \gamma.$$

Now, apply the Singular Stoke's Theorem to $G - \overline{g}$:

$$\int_{G-g} df^*(\Lambda_a) \wedge d^\perp\psi \wedge \chi = m_f(\Gamma,a) - m_f(\gamma,a) - D_f(G,a)$$

which also proves the existence of these integrals.

According to Lemma 3.3, $\psi\colon M \to \mathbb{R}$ satisfies a Lipschitz condition. The Residue Theorem can be applied to G with $T = \phi$ and $d^\perp\psi = 0$ on g:

$$T_f(G) = N_f(G,a) + \int_{G-g} d\psi \wedge d^\perp f^*(\Lambda_a) \wedge \chi .$$

Now, Remark 4 of §3 proves the theorem, q.e.d.

Now, integration over the fibers will be used. The reader is referred to the Appendix II for notations and for the properties of this operation.

Recall, that the following situation is given:

On the complex manifold N of dimension n, an admissible family of codimension s is given by the defining triplet $N \xleftarrow{\tau} F \xrightarrow{\pi} A$

where dim $A = k$ and dim $S_a = p < n$ for $a \in A$. Let t be the fiber dimension of τ. Then dim $F = n + t = k + p$ or $p = s + t$, because $s = n - p$.

Now, let ω be a non-negative form of bidegree (k,k) and of class C^∞ on A. Then

$$\Omega = \tau_*\pi^*\omega$$

is a non-negative form of bidegree (s,s) and of class C^∞ on N, where π^* denotes the pullback by π and τ_* the integration over the fibers of τ. The form Ω is <u>represented</u> by ω.

<u>Theorem 4.2. Let N, $\mathfrak{A}$, ω and Ω be given as said. Let M be a complex manifold of dimension m with $q = m - s \geq 0$. Let χ be a strictly non-negative form of bidegree (q,q) and class C^∞ on M with $d\chi = 0$ on M. Let $B = (G,\Gamma,g,\gamma,\psi)$ be a bump on M. Let $f: M \to N$ be a holomorphic map which is almost adapted to $\mathfrak{A}$. Then $N_f(G,a)$ exists and is continuous on[14] $L(\overline{G})$; moveover:</u>

$$A_f(G) = \int_A n_f(G,a)\omega$$

$$T_f(G) = \int_A N_f(G,a)\omega.$$

<u>Proof</u>. Define $\hat{f} = \pi \circ \tilde{f}: f^*(F) \to A$. Then $\psi \circ \sigma$ has compact support on $f^*(F)$ in $\sigma^{-1}(\overline{G})$. Moreover, $\hat{f}$ is open in a neighborhood of $\hat{f}^{-1}(a) \cap \sigma^{-1}(\overline{G})$ if $a \in L(\overline{G})$. Define $F(y) = \hat{f}^{-1}(y)$, then

$$N_f(G,y) = \int_{F(y)} \nu_{\hat{f}}(\psi \circ \sigma)\sigma^*(\chi)$$

exists in a neighborhood of a and is continuous at a.

Now, $\tau \circ \tilde{f} = f \circ \sigma$ implies[15] $f^* \circ \tau_* = \sigma_* \circ \tilde{f}^*$. Hence

$$\int_{f^*(F)} (\psi \circ \sigma)\sigma^*\chi \wedge \hat{f}^*(\omega)$$

$$= \int_{f^*(F)} (\psi \circ \sigma)\tilde{f}^*(\pi^*(\omega)) \wedge \sigma^*(\chi)$$

$$= \int_M \psi f^*(\tau_*\pi^*(\omega)) \wedge \chi = T_f(G)\,.$$

Let S be the set of points of $f^*(F)$ where $\hat{f}$ is not regular. According to Proposition 2.6, S is a thin analytic subset of $f^*(F)$; hence S is a set of measure zero. By Sard's theorem, also $\hat{f}(S) = S'$ is a set of measure zero on A. Appendix II Theorem AII 4.11 implies

$$\int_{f^*(F)} (\psi \circ \sigma)\sigma^*\chi \wedge \hat{f}^*(\omega)$$

$$= \int_A \left(\int_{F(a)-S} (\psi \circ \sigma)\sigma^*(\chi)\right)\omega(a)$$

$$= \int_{A-S'} \left(\int_{F(a)} \nu_{\hat{f}}(\psi \circ \sigma)\sigma^*(\chi)\right)\omega(a)$$

$$= \int_A N_f(G,a)\omega(a)\,.$$

The proof for the spherical image goes the same way; q.e.d.

If $B = (G,\Gamma,g,\gamma,\psi)$ is a bump on M, define

$$G_\sigma = \sigma^{-1}(G) \qquad \Gamma_\sigma = \sigma^{-1}(\Gamma)$$

$$g_\sigma = \sigma^{-1}(g) \qquad \gamma_\sigma = \sigma^{-1}(\gamma)$$

Then $B_\sigma = (G_\sigma, \Gamma_\sigma, g_\sigma, \gamma_\sigma, \psi \circ \sigma)$ is a bump on $f^*(F)$. Moreover, f is adapted to $\mathfrak{A}$ at x for a if and only if $\hat{f}$ is adapted to the point family $\mathfrak{A}_A$ on A at each $x' \in \sigma^{-1}(x)$ for a, because $f(x) \in S_a$ if and only if $\hat{f}(x') = a$. Hence

$$N_f(G,a) = N_{\hat{f}}(G_\sigma,a) = \int_{F(a)} v_{\hat{f}}(\psi \circ \sigma)\sigma^*(\chi)$$

$$T_f(G) = T_{\hat{f}}(G_\sigma) = \int_{G_\sigma} (\psi \circ \sigma)\hat{f}^*(\omega) \wedge \sigma^*(\chi).$$

Take $a \in A$. Suppose, that a weak proximity form λ_a for the point family $\mathfrak{A}_A$ and ω is given. Then

$$d^\perp d\lambda_a = \omega \qquad \text{on } A - \{a\}.$$

A non-negative form Λ_a of bidegree (s-1,s-1) and class C^∞ is defined by

$$\Lambda_a = \tau_* \pi^*(\lambda_a) \qquad \text{on } N - S_a.$$

with

$$d^\perp d\Lambda_a = \Omega = \tau_* \pi^*(\omega)$$

on $N - S_a$. Suppose that f is adapted to $\mathfrak{A}$ for a at every $x \in f^{-1}(S_a) \cap \overline{G}$, then $\hat{f}$ is adapted to $\mathfrak{A}_A$ for a at every $x' \in \hat{f}^{-1}(a) \cap \overline{G}_a$. Appendix II Theorem AII 4.11 , Lemma AII 5.4 and Theorem AII 4.16 imply

$$m_{\hat{f}}(\Gamma_\sigma, a) = \int_{\Gamma_\sigma} \hat{f}^*(\lambda_a) \wedge d^\perp(\psi \circ \sigma) \wedge \sigma^*(\chi)$$

$$= \int_{\Gamma_\sigma} \tilde{f}^*(\pi^*(\lambda_a)) \wedge \sigma^*(d^\perp\psi \wedge \chi)$$

$$= \int_\Gamma \sigma_* \tilde{f}^*(\pi^*(\lambda_a)) \wedge d^\perp\psi \wedge \chi$$

$$= \int_\Gamma f^*(\tau_*\pi^*(\lambda_a)) \wedge d^\perp\psi \wedge \chi$$

$$= m_f(\Gamma, a).$$

Similarly,

$$m_{\hat{f}}(\gamma_\sigma, a) = m_f(\gamma, a)$$

$$D_{\hat{f}}(G_\sigma, a) = D_f(G, a).$$

Because the First Main Theorem holds for $\hat{f}$, it also holds for f if $\Lambda_a = \tau_*\pi^*(\lambda_a)$ and $\Omega = \tau_*\pi^*(\omega)$ are used. However, that in itself does not prove that Λ_a is a weak proximity form, never the less it is true:

<u>Theorem 4.3</u>. <u>$\Lambda_a = \tau_*\pi^*(\lambda_a)$ is a weak proximity form for $\Omega = \tau_*\pi^*(\omega)$.</u>

<u>Proof</u>. Only the Residue Theorem and the Singular Stoke's

Theorem remain to be proved.

a) Suppose the assumptions of the Residue Theorem are made and assume that $\psi = 0$ on $S = \overline{H} - H$. Define $H_\sigma = \sigma^{-1}(H)$ and $S_\sigma = \sigma^{-1}(H) = \overline{H}_\sigma - H_\sigma$. Then S_σ is a boundary manifold of H_σ or empty. Because $\psi \circ \sigma | S_\sigma = 0$, the assumptions of the Residue Theorem for $\hat{f}, H_\sigma, S_\sigma, a \in A, \mathfrak{A}_A, \lambda_a, \omega, \sigma^*(\chi), \psi \circ \sigma$ are satisfied. Hence all the following integrals exist and satisfy the identity

$$\int_{H_\sigma} (\psi \circ \sigma) \hat{f}^*(\omega) \wedge \sigma^*(\chi) = \int_{H_\sigma \cap F(a)} \nu_{\hat{f}} (\psi \circ \sigma) \sigma^*(\chi)$$

$$+ \int_{H_\sigma} (\psi \circ \sigma) \hat{f}^*(\lambda_a) \wedge \sigma^*(\chi) + \int_{\overline{H}} \psi \circ \sigma d^\perp \hat{f}^*(\lambda_a) \wedge d\chi .$$

Integration over the fibers of σ implies as before the formula of the residue theorem, where the boundary integral is zero. (Appendix II Theorem A II 4.11, and Theorem A II 4.16).

b) Suppose, that the assumptions of the Singular Stoke's Theorem are made. Assume in addition that ψ is constant on each component of S and that $\psi | \overline{H}$ assumes a relative minimum or maximum on each component of S. Then the same is true for $\psi \circ \sigma$ on S_σ. Lemma 3.2 implies that $j^*((d^\perp \psi \circ \sigma) \wedge \sigma^*(\chi) \wedge f^*(\lambda_a))$ is non-negative or non-positive on each component of S_σ. Hence the assumptions for the Singular Stoke's Theorem on $f^*(F)$ for $H_\sigma, S_\sigma, \omega, \lambda_a, \psi \circ \sigma$, $\sigma^*(\chi), \hat{f}, \mathfrak{A}_A$ and $a \in A$ are satisfied:

$$\int_{H_\sigma} d\hat{f}^*(\lambda_a) \wedge d^{\perp}(\psi \circ \sigma) \wedge \sigma^*(\chi)$$

$$= \int_{S_\sigma} \hat{f}^*(\lambda_a) \wedge d^{\perp}(\psi \circ \sigma) \wedge \sigma^*(\chi) - \int_{H_\sigma} \hat{f}^*(\lambda_a) \wedge dd^{\perp}(\psi \circ \sigma) \wedge \sigma^*(\chi)$$

$$+ \int_{H_\sigma} \hat{f}^*(\lambda_a) \wedge d^{\perp}(\psi \circ \sigma) \wedge d\sigma^*(\chi)$$

Integration over the fibers of σ implies as before the formula of the Singular Stoke's Theorem. (Appendix II Theorem A II 4.11, Theorem A II 4.16 and Lemma A II 3.4).

Since the Fubini Theorem works only in one direction, the question if $\Lambda_a = \tau_* \pi^*(\lambda_a)$ is a proximity form for the form $\Omega = \tau_* \pi^*(\omega)$ remains unsettled.

§5. The construction of singular potentials

As the last theorems show, it is sufficient for many cases to construct a proximity form for the point family only. This construction shall be started now. At first some definitions and observations:

As usually, class C^k stands for "continuous" if $k = 0$ and for k-times continuously differentiable if $1 \leqq k \leqq \infty$. In addition, the following conventions will be used

$k = \mu$ measurable $\qquad$ $k = \rho$ real analytic

$k = \beta$ locally bounded and measurable, $\qquad$ $k = \omega$ holomorphic

$k = k_1 \cap k_2$ of class C^{k_1} and of class C^{k_2}

If functions, maps, etc., are considered on a product of two manifolds, $C^k(u,v)$ stands for class C^k in both variables, class C^u in the first variable for each fixed value of the second variable, and class C^v in the second variable for each fixed value of the first variable.

Let $\pi: E \to M$ be a fiber bundle over M. Then $\Gamma(U,E)$ denotes the set of sections in E over the subset U of E. Write also $\Gamma(E) = \Gamma(M,E)$. Moreover, $\Gamma^k(U,E) = \Gamma(U,E,C^k)$ denotes the set of sections of class C^k in E over U, provided this concept applies.

Let N be a complex manifold of dimension n. Let $T = T(N)$ be the complexified cotangent bundle. Let $S = S(N)$ be the holomorphic cotangent bundle and $\overline{S} = \overline{S}(N)$ be its conjugate. Then $T = S \oplus \overline{S}$. Define

$$\left.\begin{aligned} T^p &= T^p(N) = T \wedge \cdots \wedge T \\ S^p &= S^p(N) = S \wedge \cdots \wedge S \end{aligned}\right\} \quad \text{(p-times)}$$

$$T^{p,q} = T^{p,q}(N) = S^p \wedge \bar{S}^q$$

Then $T^m = \bigoplus_{p+q=m} T^{p,q}$.

Let A be a differentiable or complex manifold and let $\tau\colon N \times A \to N$ be the natural projection. For each $t \in A$, the submanifold $N \times \{t\}$ is a complex manifold and $\tau_t = \tau|(N \times \{t\})$ maps $N \times \{t\}$ biholomorphically onto N. Let $j_t\colon N \times \{t\} \to N \times A$ be the inclusion map. Let

$$\hat{T}^{pq} = \hat{T}^{pq}(N) = \tau^*(T^{pq}(N))$$

$$\hat{T}^m = \hat{T}^m(N) = \tau^*(T^m(N)) = \bigoplus_{p+q=m} \hat{T}^{pq}$$

be the pull back of the vector bundles which can and will be regarded as subbundles of $T^{p+q}(N \times A)$ resp. $T^m(N \times A)$. Each section $s \in \Gamma(U, T^{p,q})$ lifts to a section $\tau^*(s) \in \Gamma(\tau^{-1}(U), \hat{T}^{p,q})$ and similarly of T^m.

If U is open in $N \times A$, then $U_t = \{x \,|\, (x,t) \in U\}$ is open in N. If $\chi \in \Gamma(U, \hat{T}^m)$ and if $t \in A$ with $U_t \neq \emptyset$, define

$$\chi_t = (j_t \circ \tau_t^{-1})^*(\chi) \in \Gamma\,(U_t, T^m)$$

If $\chi \in \Gamma(U,\hat{T}^{p,q})$ then $\chi_t \in \Gamma(U_t,T^{p,q})$ is a form of bidegree (p,q).

Let $\alpha: W \to W'$ be a biholomorphic map of an open subset W of N onto an open subset W' of $\mathbb{C}^n$. Then $\alpha = (z_1,\ldots,z_n)$ and $\{dz_\mu \wedge d\bar{z}_\nu | (\mu,\nu) \in T(p,n) \times T(q,n)\}$ is a frame field of $T^{p,q}$ over W. Then $\tau^*(dz_\mu) = d(z_\mu \circ \tau)$ and $\{\tau^*(dz_\mu) \wedge \tau^*(d\bar{z}_\nu) | (\mu,\nu) \in T(p,n) \times T(q,n)\}$ is a frame field of $\hat{T}^{p,q}$ over $\tau^{-1}(W) = V$. Moreover, $\chi \in \Gamma^k(V,\hat{T}^{p,q})$ if and only if

$$\chi = \sum_{\mu\in T(p,n)} \sum_{\nu\in T(q,n)} \chi_{\mu\nu}\tau^*(dz_\mu) \wedge \tau^*(d\bar{z}_\nu)$$

and if the function $\chi_{\mu\nu}$ are of class C^k on U. If $t \in N$, then

$$\chi_t(x) = \sum_{\mu\in T(p,n)} \sum_{\nu\in T(q,n)} \chi_{\mu\nu}(x,t)dz_\mu \wedge d\bar{z}_\nu$$

for $x \in W$. If $\chi = \tau^*(\varphi)$ then $\tau^*(\varphi)_t = \varphi$.

Take $\Lambda = N$. Let $\tau: N \times N \to N$ be the projection onto the first factor and $\pi: N \times N \to N$ onto the second factor. Let $F_N = \{(x,x) | x \in N\}$ be the diagonal. Then $N \xleftarrow{\tau} F_N \xrightarrow{\pi} N$ defines the point family $\mathfrak{A}_N$, which will be studied first. Form $\hat{T}^m$ and $\hat{T}^{p,q}$ in respect to the projection τ. Define $\hat{N} = N \times N - F_N$.

<u>Definition 5.1.</u> <u>Let $\omega \in \Gamma^\infty(N \times N,\hat{T}^{n,n})$. Then $\lambda \in \Gamma^\infty(\hat{N},\hat{T}^{n-1,n-1})$ is said to be a singular potential on N for ω if and only if the following conditions are satisfied:</u>

1) For every $t \in N$, the identity $d^{\perp}d\lambda_t = \omega_t$ holds on $N - \{t\}$. Moreover, $\lambda_t \geqq 0$ on $N - \{t\}$ for each $t \in N$.

2) If $a \in N$, if $\alpha: U \to U'$ is a biholomorphic map of an open neighborhood U of a onto an open subset U' of $\mathbb{C}^n$, then an open, relative compact neighborhood V of a with $\overline{V} \subseteq U$, a non-negative integer κ and locally bounded forms $\eta \in \Gamma^{\beta}(V \times V, \hat{T}^{n-1,n-1})$ and $\sigma \in \Gamma^{\beta}(V \times V, \hat{T}^{2n-1})$ exist such that

$$\lambda(x,t) = \frac{(\log|\alpha(x)-\alpha(t)|)^{\kappa}}{|\alpha(x)-\alpha(t)|^{2n-2}}\eta(x,t) \qquad \text{on } V \times V - F_N$$

$$d^{\perp}\lambda_t(x) = \frac{(\log|\alpha(x)-\alpha(t)|)^{\kappa}}{|\alpha(x)-\alpha(t)|^{2n-1}}\sigma_t(x) \qquad \text{on } V - \{t\} \text{ all } t \in V$$

3) If $a \in N$, then an open neighborhood W of a and biholomorphic maps $\alpha_j: W \to W'_j$ onto open subsets W'_j of $\mathbb{C}^n$ with $\alpha_j(a) = 0$ exist for $j = 1,\ldots,r$ such that [16)]

$$d^{\perp}\lambda_a(x) = \frac{(n-1)!}{2\pi^n}\sum_{j=1}^{r}\mu_j(x)d^{\perp}\log\frac{1}{|\alpha_j(x)|} \wedge \frac{\alpha_j^*(\upsilon_{n-1})(x)}{|\alpha_j(x)|^{2n-2}}$$

$$+ \sum_{j=1}^{r}(\log|\alpha_j(x)|)^{\kappa_j}\frac{\rho_j(x)}{|\alpha_j(x)|^{2n-2}}$$

for $x \in W - \{a\}$. Here each κ_j is a non-negative integer. Each $\mu_j: W \to \mathbb{R}$ is continuous with

$$\sum_{j=1}^{r} \mu_j(a) = 1 .$$

Moreover, each ρ_j is locally bounded on W and of class C^∞ on W - {a}:

$$\rho_j \in \Gamma^\beta(W,T^{2n-1}) \cap \Gamma^\infty(W-\{a\},T^{2n-1}) .$$

Hirschfelder [7a] has shown that the proximity form of Wu [33] is a singular potential. Also the Levine form [28], p.150 is a singular potential. In both cases r = n and p = 0. Probably the proximity form of Chern [3] is related to a singular potential.

Lemma 5.2. Let $N \neq \emptyset$ be an open subset of $\mathbb{C}^n$. If $n > 1$ define

$$\lambda(x,t) = \frac{(r-2)!}{4\pi^n} \frac{1}{|x-t|^{2n-2}} \upsilon_{n-1}(x)$$

for $(x,t) \in \hat{N}$. If n = 1, assume that N has a finite diameter $D > 0$. Define

$$\lambda(x,t) = \frac{1}{2\pi} \log \frac{D}{|x-t|}$$

Then λ is a singular potential for 0.

Proof. Obviously $\lambda_t \geqq 0$ on N - {t} for each $t \in N$. Moreover,

$$d^{\perp}\lambda_t(x) = -\frac{i(n-1)!}{4\pi^n}\frac{\partial|x-t|^2 - \bar{\partial}|x-t|^2}{|x-t|^{2n}} \wedge v_{n-1}(x)$$

$$= \frac{(n-1)!}{2\pi^n}\, d^{\perp}\log\frac{1}{|x-t|} \wedge \frac{v_{n-1}(x)}{|x-t|^{2n-2}}$$

on $N - \{t\}$ for $n \geqq 1$. Hence, condition 3 of Definition 3.1 is satisfied. Now

$$dd^{\perp}\lambda_t(x) = \frac{i(n-1)!}{2\pi^n}\frac{\partial\bar{\partial}|x-t|^2}{|x-t|^{2n}} \wedge v_{n-1}(x)$$

$$-\frac{in!}{2\pi^n}\frac{\partial|x-t|^2 \wedge \bar{\partial}|x-t|^2}{|x-t|^{2n+2}}\, v_{n-1}(x)$$

where

$$\tfrac{i}{2}\partial|x-t|^2 \wedge \bar{\partial}|x-t|^2 \wedge v_{n-1}(x)$$

$$= \left(\tfrac{i}{2}\right)^n \sum_{\mu,\nu=1}^{n} (\bar{x}_\mu-\bar{t}_\mu)(x_\nu-t_\nu)dx_\mu \wedge d\bar{x}_\nu \wedge \sum_{\delta=1}^{n} \prod_{\substack{\lambda=1\\ \lambda\neq\delta}}^{n} dx_\mu \wedge d\bar{x}_\mu$$

$$= \left(\tfrac{i}{2}\right)^n \sum_{\delta=1}^{n} (\bar{x}_\delta-\bar{t}_\delta)(x_\delta-t_\delta)dx_1 \wedge d\bar{x}_1 \wedge \dots \wedge dx_n \wedge d\bar{x}_n$$

$$= |x-t|^2\, v_n(x)$$

$$\tfrac{i}{2}\,\partial\bar{\partial}|x-t|^2 \wedge v_{n-1}(x) = \tfrac{i}{2}\,\partial\bar{\partial}|x|^2 \wedge v_{n-1}(x) = \tfrac{i}{2}\, n v_n(x)$$

Hence $dd^{\perp}\lambda_t(x) = 0$ for $x \in N - \{t\}$, which proves condition 1.

Take $a \in N$. Let $\alpha: U \to U'$ be a biholomorphic map of an open neighborhood U of a onto an open subset U' of $\mathbb{C}^n$. Let V be an open neighborhood of a with $\overline{V} \subset U$ where $\overline{V}$ is compact and such that diam $\alpha(V) < 1$. Then positive constants C_1 and C_2 exist such that

$$C_1|x-t| \leqq |\alpha(x)-\alpha(t)| \leqq C_2|x-t|$$

for all $(x,t) \in V \times V$. Define $\sigma \in \Gamma^{\beta}(V{\times}V,\hat{T}^{2n-1})$ by

$$\sigma(x,t) = i\frac{(n-1)!}{4\pi^n}\left(\frac{|\alpha(x)-\alpha(t)|}{|x-t|}\right)^{2n-1} \frac{\overline{\partial}|x-t|^2-\partial|x-t|^2}{|x-t|} \wedge \upsilon_{n-1}(x)$$

for $(x,t) \in V \times V - F_N$ and $\sigma(x,x) = 0$ if $x \in V$. Define $\eta \in \Gamma^{\beta}(V{\times}V_1\hat{T}^{2n-2})$ for $n > 1$ by

$$\eta(x,t) = \frac{(n-2)!}{4\pi^n}\left(\frac{|\alpha(x)-\alpha(t)|}{|x-t|}\right)^{2n-2} \upsilon_{n-1}(x)$$

for $(x,t) \in V \times V - F_N$ and for $n = 1$ by

$$\eta(x,t) = \frac{1}{2\pi}\left(\log \frac{D|\alpha(x)-\alpha(t)|}{|x-t|}\right) \frac{1}{\log|\alpha(x)-\alpha(t)|} - \frac{1}{2\pi}$$

for $(x,t) \in V \times V - F_N$. Define $\beta(x,x) = 0$. With these definitions condition 2 is satisfied, q.e.d.

Hence a singular potential for some ω can be constructed locally. Now, a partition of unity is used, to construct a global solution for some ω:

<u>Theorem 5.3.</u> <u>Let N be a complex manifold of dimension n. Then</u>

<u>a singular potential λ for some form ω exists where $\omega | G = 0$ on some neighborhood G of the diagonal F_N of N x N.</u>

<u>Proof</u>. According to Lemma 5.2, open, locally finite coverings $\{W_\mu\}_{\mu \in M}$ and $\{Z_\mu\}_{\mu \in M}$ of N with $\emptyset \neq Z_\mu \subset \overline{Z}_\mu \subset W_\mu$ and with $\overline{Z}_\mu$ compact and a family $\{\lambda^\mu\}_{\mu \in M}$ of singular potentials λ^μ on W_μ for O exist. Take a partition of unity $\{g_\mu\}_{\mu \in M}$ by non-negative C^∞-functions on N such that $G_\mu = \operatorname{supp} g_\mu \subset Z_\mu$ and $\sum_{\mu \in M} g_\mu = 1$ on N. Observe that $W = \bigcup_{\mu \in M} W_\mu \times W_\mu$ is an open neighborhood of the diagonal F_N. For $\mu \in M$, take a non-negative C^∞-function h_μ on N with compact support $H_\mu \subset W_\mu$ such that $h_\mu | \overline{Z}_\mu = 1$. Let τ: N x N $\to$ N be the projection onto the first factor and let π: N x N $\to$ N be the projection onto the second factor. Then

$$T_\mu = \operatorname{supp} [(g_\mu \circ \pi)\cdot(h_\mu \circ \tau)] = H_\mu \times G_\mu \subseteq W_\mu \times W_\mu$$

where T_μ is compact. On $W_\mu \times W_\mu - F_N$ define $\tilde{\lambda}^\mu = (g_\mu \circ \pi)(h_\mu \circ \tau)\lambda^\mu$. On the complement $N - W_\mu \times W_\mu$, define $\tilde{\lambda}^\mu = 0$. Then $\lambda = \sum_{\mu \in M} \tilde{\lambda}^\mu \in \Gamma^\infty(\hat{N}, \hat{T}^{n-1,n-1})$ where $\hat{N} = N \times N - F_N$. Also, $\omega \in \Gamma^\infty(\hat{N}, \hat{T}^{n,n})$ is well defined by $\omega_t = d^\perp d\lambda_t$ on $N - \{t\}$ for all $t \in N$. Define $\omega(x,x) = 0$ for all $x \in N$.

Now, it shall be shown, that $\omega = 0$ in a neighborhood of F_N. Pick $(a,a) \in F_N$. Define $M(a) = \{\mu | (a,a) \in T_\mu\}$. Then M(a) is finite. Define

$$Z^a = \bigcap_{\mu \in M(a)} Z_\mu \times Z_\mu$$

$$T^a = \bigcup_{\nu \in M - M(a)} T_\nu .$$

If $(b,c) \in \overline{T^a}$, pick a neighborhood V of c such that $\{\mu \in M | V \cap W_\mu \neq \emptyset\}$ if finite. A sequence $(x_\lambda, y_\lambda) \in T^a$ exists such that $(x_\lambda, y_\lambda) \to (b,c)$ for $\lambda \to \infty$. Then $(x_\lambda, y_\lambda) \in T_{\nu_\lambda}$ with $\nu_\lambda \in M - M(a)$. Hence $y_\lambda \in G_{\nu_\lambda} \subseteq W_{\nu_\lambda}$. A number λ_0 exists such that $y_\lambda \in V$ for all $\lambda \geqq \lambda_0$. Hence $V \cap W_{\nu_\lambda} \neq \emptyset$ for $\lambda \geqq \lambda_0$. Hence $\nu \in M - M(a)$ exists such that $\nu = \nu_\lambda$ for infinitely many λ. Hence $(b,c) \in T_\nu \subseteq T^a$. Therefore T^a is closed and does not contain a. If $\mu \in M(a)$, then $(a,a) \in T_\mu$ and $a \in G_\mu \subseteq Z_\mu$. Hence $Z^a - T^a$ is an open neighborhood of (a,a). Therefore, an open neighborhood V_a of a in N exists such that $V_a \times V_a \subseteq Z^a - T^a$. If $(x,t) \in V_a \times V_a$, then $x \in Z_\mu$ and $h_\mu(x) = 1$ for all $\mu \in M(a)$. Consequently,

$$\lambda(x,t) = \sum_{\mu \in M(a)} \tilde{\lambda}^\mu(x,t) = \sum_{\mu \in M(a)} g_\mu(t) \lambda^\mu(x,t)$$

for $(x,t) \in V_a \times V_a - F_N$. Hence

$$\omega(x,t) = d^\perp d\lambda_t(x) = \sum_{\mu \in M(a)} g_\mu(t) d^\perp d\lambda_t^\mu(x) = 0$$

on $V_a \times V_a - F_N$. Now, $G = \bigcup_{a \in N} V_a \times V_a$ is an open neighborhood of F_N

with $\omega|G = 0$. This proves that ω is of class C^∞ on $N \times N$.

Now, ω and λ are defined and the conditions of definition 5.1 have to be proved. By construction $\lambda_t \geqq 0$ and $d^\perp d\lambda_t = \omega_t$ on $N - \{t\}$. Therefore, condition 1 holds.

Take $a \in N$. Let $\alpha\colon U \to U'$ be a biholomorphic map of an open neighborhood U of a onto an open subset U' of $\mathbb{C}^n$. For each $\mu \in M(a)$ an open neighborhood Y_μ of a with $Y_\mu \subseteq \overline{Y}_\mu \subset V_a \cap U$ exists such that

$$\lambda^\mu(x,t) = \frac{(|\log|\alpha(x)-\alpha(t)||)^{\kappa_\mu}}{|\alpha(x)-\alpha(t)|^{2n-2}}\,\eta^\mu(x,t)$$

$$d^\perp\lambda^\mu_t(x) = \frac{(\log|\alpha(x)-\alpha(t)|)^{\kappa_\mu}}{|\alpha(x)-\alpha(t)|^{2n-1}}\,\sigma^\mu_t(x)$$

for $(x,t) \in Y_\mu \times Y_\mu - F_N$ where $\kappa_\mu \geqq 0$ is an integer and where $\eta^\mu \in \Gamma^\beta(Y_\mu x Y_\mu, \hat{T}^{n-1,n-1})$ and $\sigma^\mu \in \Gamma^\beta(Y_\mu x Y_\mu, \hat{T}^{2n-1})$. Because $M(a)$ is finite, $\kappa = \kappa_\mu$ can be taken independently of μ. Define $V = \bigcap_{\mu \in M(a)} Y_\mu$. Then $a \in V \subset \overline{V} \subset U \cap V_a$ and $\overline{V}$ is compact. Now condition 2 is satisfied on V with

$$\eta(x,t) = \sum_{\mu \in M(a)} g_\mu(t)\eta^\mu(x,t)$$

$$\sigma(x,t) = \sum_{\mu \in M(a)} g_\mu(t)\sigma^\mu(x,t)\,.$$

Take $a \in N$. Then

$$\lambda_a(x) = \sum_{\nu \in M(a)} g_\nu(a)\lambda_a^\nu(x)$$

for $x \in V_a - \{a\}$. Hence

$$d^\perp\lambda_a(x) = \sum_{\nu \in M(a)} g_\nu(a) d^\perp\lambda_a^\nu(x)$$

for $x \in V_a - \{a\}$.

Take $\nu \in M(a)$. Then an open neighborhood X_ν of a and biholomorphic maps $\alpha_\nu : X_\nu \to X'_{\nu j}$ onto an open subset $X'_{\nu j}$ of $\mathbb{C}^n$ with $\alpha_{\nu j}(0) = 0$ exist for $j = 1,\ldots,r(\nu)$ such that

$$d^\perp\lambda_a^\nu(x) = \frac{(n-1)!}{2\pi^n}\sum_{j=1}^{r(\nu)} \mu_{\nu j}(x) d^\perp \log \frac{1}{|\alpha_{\nu j}(x)|} \wedge \frac{\alpha_{\nu j}^*(\upsilon_{n-1})(x)}{|\alpha_{\nu j}(x)|^{2n-2}}$$

$$+ \sum_{j=1}^{r(\nu)} (\log|\alpha_{\nu j}(x)|)^{\kappa_{\nu j}} \frac{\rho_{\nu j}(x)}{|\alpha_{\nu j}(x)|^{2n-2}}$$

for $x \in X_\nu - \{a\}$. Each $\mu_{\nu j}: X_\nu \to \mathbb{R}$ is continuous with

$$\sum_{j=1}^{r(\nu)} \mu_{\nu j}(a) = 1$$

Each $\kappa_{\nu j}$ is a non-negative integer and $\rho_{\nu j}$ is a form of degree $2n-1$, which is locally bounded on X_ν and of class C^∞ on $X_\nu - \{a\}$. Restrict to the open neighborhood $X = V_a \cap \bigcap_{\nu \in M(a)} X_\nu$ of a. Then

$$1 = \sum_{\nu \in M} g_\nu(a) = \sum_{\nu \in M(a)} g_\nu(a)$$

implies

$$\sum_{\nu \in M(a)} \sum_{j=1}^{r(\nu)} g_\nu(a)\mu_{\nu j}(a) = 1$$

Therefore, condition 3 is satisfied with $\alpha_{\nu j} : X \to \alpha_{\nu j}(X)$, with $g_\nu(a)\mu_{\nu j}$, with $\kappa_{\nu j}$ and $g_\nu(a)\rho_{\nu j}$; q.e.d.

A singular potential has been constructed for some ω. Later, a singular potential for a given ω shall be constructed, namely, if ω is the volume element of a Kaehler manifold. However, it is more convenient, if not absolutely necessary, to explore at first some of the properties of a singular potential, namely: λ_t is a proximity form for ω_t for each $t \in N$.

§6. Properties of singular potentials

It will be shown, that certain integrals involving a singular potential depend continuously on t and that a singular potential λ defines a proximity form λ_t for the point family for each $t \in N$.

Recall the convention of "class" on product manifolds. For instance, on a subset of a product of two spaces χ is of class $C^\beta(\mu,0)$, if χ has measurable and locally bounded coefficients, if each coefficient is measurable in the first variable, for each fixed value of the second and if χ is continuous in the second variable, for each fixed value of the first.

Proposition 6.1. Let M and N be complex manifolds with $\dim M = m$ and $\dim N = n$. Suppose that $q = m - n \geqq 0$. Let χ be a form of bidegree $(q+1,q+1)$ and of class $C^\beta(\mu;0)$ on $M \times N$. For $t \in N$, let $j_t\colon M \to M \times N$ be defined by $j_t(z) = (z,t)$. Let $f\colon M \to N$ be a holomorphic map. Let K be a measurable, relative compact subset of M. Let $N(\overline{K})$ be the set of all $t \in N$ such that $f|U_x$ is open for some neighborhood U_x of x for every $x \in f^{-1}(t) \cap \overline{K}$ or such that $f^{-1}(t) \cap \overline{K}$ is empty. Then $N(\overline{K})$ is open.

Let $\lambda \in \Gamma^\infty(\hat{N},\hat{T}^{n-1,n-1})$ be a singular potential on N for some $\omega \in \Gamma^\infty(N\times N,\hat{T}^{n,n})$. Then the integral

$$F(t) = \int_K f^*(\lambda_t) \wedge j_t^*(\chi)$$

exists for each point $t \in N(\overline{K})$. Moreover F is continuous on $N(\overline{K})$.

Remark: Suppose that the assumptions of the proposition are made with the exception of those about χ. However, let χ be a locally bounded and measurable form of bidegree $(q+1,q+1)$ on M. Let $\pi\colon M \times N \to M$ be the projection. Then the theorem applies to $\pi^*(\chi)$. Moreover, $j_t^*\pi^*(\chi) = \chi$ for each $t \in N$. Therefore

$$\int_K f^*(\lambda_t) \wedge \chi$$

exists for $t \in N(\overline{K})$ and defines a continuous function on $N(\overline{K})$.

Proof. Because $\overline{K}$ is compact, $N(\overline{K})$ is open as easily seen.[17)] Take any $a \in N(\overline{K})$. It has to be proved that F is continuous in a neighborhood of a. Pick $z_0 \in \overline{K}$. Then the following statement shall be proved:

Statement. An open neighborhood M_0 of z_0 and an open neighborhood N_0 of a exist such that, for every C^∞-function $\mu\colon M \to \mathbb{R}$ with compact support in M_0, the integral

$$F_\mu(t) = \int_K f^*(\lambda_t) \wedge \mu j_t^*(\chi)$$

exists for each $t \in N_0$ and defines a continuous function F_μ on N_0.

Suppose that the Statement is proved: Then finitely many of these neighborhoods $M_0^1,\ldots,M_0^r$ cover K. Take C^∞-functions $\mu_\delta \geqq 0$ such that $\sum_{\delta=1}^{r} \mu_\delta = 1$ on $\overline{K}$ and such that $\operatorname{supp} \mu_\delta \subset M_0^\delta$ for $\delta = 1,\ldots,r$.

Then $F = \sum_{\delta=1}^{r} F_{\mu_\delta}$ is continuous on the neighborhood $N_0^1 \cap \ldots \cap N_0^r$ of a.

It remains to prove the Statement: If $f(z_0) \neq a$ take M_0 as an open, relative compact neighborhood of z_0 and N_0 as an open, relative compact neighborhood of a such that $f(\overline{M}_0) \cap \overline{N}_0 = \emptyset$. Obviously, the Statement is true.

Therefore, only the case $f(z_0) = a$ has to be considered. Pick $\alpha: U \to U'$, V, η and κ such that condition 2) of Definition 5.1 holds. Moreover, U can be taken so small that $f|U$ is open. Define $N_0 = V$ and $M_0 = f^{-1}(V) \cap U$. Suppose that the C^∞-function μ on M has compact support in M_0. Define $K_1 = \overline{K} \cap \operatorname{supp} \mu$. Identify $U = U'$ such that α becomes the identity. Define $\rho(z) = 1$ if $z \in \overline{K} - K$ and $\rho(z) = 0$ if $z \in M - (\overline{K} - K)$. Appendix I Theorem A I 10 implies that for every $t \in N_0$ the integral

$$\int_{z \in K_1} \left(\log \frac{1}{|f(z)-t|}\right)^k \frac{1}{|f(z)-t|^{2n-2}} f^*(\eta_t(z)) \wedge \mu(z)\rho(z) j_t^*(\chi)$$

$$= \int_K f^*(\lambda_t) \wedge \mu j_t^*(\chi) = F_\mu(t)$$

exists and is a continuous function of t on N_0, q.e.d.

Proposition 6.2. Let M and N be complex manifolds with dim M = m and dim N = n. Suppose that $q = m - n \geqq 0$. Let χ be a form of degree $2q + 1$ and of class $C^\beta(\mu;0)$ on M x N. For $t \in N$, define $j_t: M \to M \times N$ by $j_t(z) = (z,t)$. Let $f: M \to N$ be a holomorphic map.

Let K be a measurable, relative compact subset of M. Let $N(\overline{K})$ be the set of all $t \in N$ such that $f^{-1}(t) \cap \overline{K}$ is either empty or each of its points $x \in f^{-1}(t) \cap \overline{K}$ has an open neighborhood U_x such that $f|U_x$ is open. Then $N(\overline{K})$ is open.

Let $\lambda \in \Gamma^{\infty}(\hat{N}, \hat{T}^{n-1,n-1})$ be a singular potential on N for some $\omega \in \Gamma^{\infty}(N\times N, \hat{T}^{n,n})$. Then the integral

$$F(t) = \int_K d^{\perp}f^*(\lambda_t) \wedge j_t^*(\chi)$$

exists for each point $t \in N(\overline{K})$. Moreover, F is continuous on $N(\overline{K})$.

Remark. Suppose that the assumptions of the proposition are made with the exception of those about χ. However, let χ be a locally bounded and measurable form of degree 2q + 1 on M. Let $\pi: M \times N \to M$ be the projection. Then the theorem applies to $\pi^*(\chi)$. Moreover, $j_t^*\pi^*(\chi) = \chi$ for each $t \in N$. Therefore

$$\int_K d^{\perp}f^*(\lambda_t) \wedge \chi$$

exists for each $t \in N(\overline{K})$ and defines a continuous function on $N(\overline{K})$.

The proof is almost the same as the proof of Proposition 6.1. Only χ has degree 2q + 1 now, and λ_t and η have to be replaced by $d^{\perp}\lambda_t$ and σ (Definition 5.1) observing that $f^*(d^{\perp}\lambda_t) = d^{\perp}f^*(\lambda_t)$.

The theory of Appendix A II §6 will be used to compute the integral average of λ_t over $t \in N$. Since t is the second variable, the factors in A II §6 have to be exchanged. Because, both factors

F and N are complex manifolds here, this causes no difficulties; F x N and N x F are biholomorphically equivalent.

Let N and A be complex manifolds of dimension n and k respectively. Let $\tau: N \times A \to N$ and $\pi: N \times A \to A$ be the projections. Let ψ be a form of degree 2k on A. Let W be an open subset of N x A. A section $\varphi \in \Gamma(W, \hat{T}^{p,q}(N))$ can be regarded as a form of type (p,q,0) on N x A in the sense of A II §6 (observe the exchange of the factors). Define φ on all of N x A by setting $\varphi = 0$ on $N \times A - W$. Then the integral average $L_\psi(\varphi)$ is defined whenever it exists. For $x \in N$, define

$$W^x = \{y \in A \mid (x,y) \in W\}.$$

Take any $a \in N$. Let $\alpha: U \to U'$ be a biholomorphic map of an open neighborhood U of a onto an open neighborhood U' of $\mathbb{C}^n$. Define $V = (U \times A) \cap W$. Set $\alpha = (z_1, \ldots, z_n)$. Then

$$\varphi = \sum_{\rho \in T(p,n)} \sum_{\sigma \in T(a,n)} \varphi_{\rho\sigma}\, \tau^*(dz_\rho \wedge d\bar{z}_\sigma)$$

on V where $\varphi_{\rho\sigma}$ are functions on V. Then $L_\psi(\varphi)$ exists at $x \in U$, if and only if all $\varphi_{\rho\sigma}(x,\cdot)\psi$ are integrable over W^x and

$$L_\psi(\varphi)(x) = \sum_{\rho \in T(p,n)} \sum_{\sigma \in T(q,n)} \left(\int_{y \in W^x} \varphi_{\rho\sigma}(x,y)\psi(y) \right) dz_\rho \wedge d\bar{z}_\sigma.$$

In the following application A = N; never the less it will be important to distinguish between first and second factor.

Proposition 6.3. Let N be a complex manifold of dimension n. Let ψ be a continuous form of degree 2n on N with compact support in N. Let $\lambda \in \Gamma^\infty(\hat{N}, \hat{T}^{n-1,n-1})$ be a singular potential on N for $\omega \in \Gamma^\infty(N\times N, \hat{T}^{2n})$. Then, for every $x \in N$, the integral average $L_\psi(\lambda)(x)$ exists and defines a continuous form $L_\psi(\lambda)$ of bidegree (n-1,n-1) on N. If $\psi \geqq 0$, then $L_\psi(\lambda) \geqq 0$.

Proof. Take $a \in N$. Take $\alpha: U \to U'$, V, κ and η such as in condition 2 of Definition 5.1. Define $V' = \alpha(V)$ and $\tilde{V} = V \times N$. Denote $\alpha = (z_1,\ldots,z_n)$. On $\tilde{V}$

$$\lambda(x,y) = \sum_{\rho,\sigma\in T(n-1,n)} \lambda_{\rho\sigma}(x,y)\tau^*(dz_\rho \wedge d\bar{z}_\sigma)$$

where $\lambda_{\rho\sigma}$ are C^∞-functions on $\tilde{V} - F_N$ and where F_N is the diagonal of N. On V x V

$$\eta(x,y) = \sum_{\rho,\sigma\in T(n-1,n)} \eta_{\rho,\sigma}(x,y)\tau^*(dz_\rho \wedge d\bar{z}_\sigma)$$

where $\eta_{\rho\sigma}$ are locally bounded functions on V x V. Then

$$\lambda_{\rho\sigma}(x,y) = \frac{(\log|\alpha(x)-\alpha(y)|)^\kappa}{|\alpha(x)-\alpha(y)|^{2n-2}}\eta_{\rho\sigma}(x,y)$$

on $V \times V - F_N$. Then

$$L_\psi(\lambda)(x) = \sum_{\rho,\sigma\in T(n-1,n)} \left(\int_{y\in N} \lambda_{\rho\sigma}(x,y)\psi(y)\right)dz_\rho \wedge d\bar{z}_\sigma$$

for $x \in V$, provided the integral exists.

Let V_1 be an open, relative compact neighborhood of a with $\overline{V}_1 \subset V$. Let μ be a C^∞-function on N with compact support K in V such that $\mu|V_1 = 1$. Let V_2 be an open neighborhood of a with $\overline{V}_2 \subset V_1$. Then $(1-\mu(y))\lambda(x,y)$ in C^∞ is a neighborhood of $\overline{V}_2 \times N$. Therefore, $L_{(1-\mu)\psi}(\lambda)$ is a C^∞-form of bidegree (n-1,n-1) on V_2. On V

$$\psi = \left(\tfrac{i}{2}\right)^n g\, dz_1 \wedge d\overline{z}_1 \wedge \cdots \wedge dz_n \wedge d\overline{z}_n$$

Define

$$\tilde{\varphi} = \left(\tfrac{i}{2}\right)^{n-1} \mu g\, dz_1 \wedge d\overline{z}_2 \wedge \cdots \wedge dz_{n-1} \wedge d\overline{z}_{n-1}$$

on V. On V', set $\varphi = (\alpha^{-1})^*(\tilde{\varphi})$. On V x V, define

$$\chi_{\rho\sigma}(x,y) = \tfrac{i}{2}\, \eta_{\rho\sigma}(x,t)\, \pi^*(dz_n \wedge d\overline{z}_n)$$

Appendix Theorem A I 10 implies that the integral

$$F_{\rho\sigma}(x) = \int_{y\in K} \left(\log \frac{1}{|\alpha(x)-\alpha(y)|}\right)^{\kappa} \frac{1}{|\alpha(x)-\alpha(y)|^{2n-2}} \alpha^*(\varphi) \wedge \chi_{\rho\sigma}(x,y)$$

$$= \int_{y\in V} \left(\log \frac{1}{|\alpha(x)-\alpha(y)|}\right)^{\kappa} \frac{1}{|\alpha(x)-\alpha(y)|^{2n-2}} \eta_{\rho\sigma}(x,y)\mu(y)\psi(y)$$

$$= \int_{y\in V} \lambda_{\rho\sigma}(x,y)\mu(y)\psi(y)$$

$$= \int_{y\in N} \lambda_{\rho\sigma}(x,y)\mu(y)\psi(y)$$

exists for $x \in V$ and defines a continuous function on V. Therefore,

$$L_{\mu\psi}(\lambda)(x) = \sum_{\rho\sigma\, T(n-1,n)} F_{\rho\sigma}(x)\, dz_\rho \wedge d\bar{z}_\sigma$$

exists for every $x \in V$ and defines a continuous function on V. Therefore,

$$L_\psi(\lambda)(x) = L_{(1-\mu)\psi}(\lambda)(x) + L_{\mu\psi}(\lambda)(x)$$

exists for $x \in V_2$ and defines a continuous function on the neighborhood V_2 of a.

For $t \in N$, define $i_t : N \to N \times N$ by $i_t(z) = (z,t)$ for each $z \in N$. Then $i_t^*(\lambda) = \lambda_t \geqq 0$ on $N - \{t\}$. Suppose that $\psi \geqq 0$ on N. Appendix II Lemma A II 6.8 implies $L_\psi(\lambda) \geqq 0$; q.e.d.

Define $P = \{\delta \in \mathbb{R} \mid 0 < \delta < 1\}$. Then $g = \{g_\rho\}_{\rho\in P}$ is called a test family[18] if the following conditions are satisfied:

1. For each $\rho \in P$, the function $g_\rho : \mathbb{R} \to \mathbb{R}$ is of class C^∞
2. If $\rho \in P$ and $x \in \mathbb{R}$, then $0 \leqq g_\rho(x) \leqq 1$
3. If $\rho \in P$ and $x \leqq \frac{\rho}{2}$, then $g_\rho(x) = 0$
4. If $\rho \in P$ and $x \geqq \rho$, then $g_\rho(x) = 1$
5. A constant $B > 0$ exists such that $\rho|g_\rho'(x)| \leqq B$ for all $x \in \mathbb{R}$ and $\rho \in B$.

Lemma A I 13 gives the existence of a test family.

A singular potential λ for ω defines a proximity form λ_t for ω_t for each $t \in N$, as shall be shown now:

<u>Theorem 6.4. Let N be a complex manifold of dimension n. Let $\lambda \in \Gamma^\infty(\hat{N}, \hat{T}^{n-1,n-1})$ be a singular potential on N for $\omega \in \Gamma^\infty(N\times N, \hat{T}^{n,n})$. Then λ_t is a proximity form of ω_t for every $t \in N$ in respect to the point family $\mathfrak{A}_N$ of N.</u>

<u>Proof</u>. By definition, $\lambda_t \geqq 0$ is a C^∞-form of bidegree $(n-1,n-1)$ on $N-\{t\}$ such that $d^\perp d\lambda_t = \omega_t$ on $N-\{t\}$. Hence, it remains to be shown that the "Residue Theorem" and the "Singular Stoke's Theorem" hold.

<u>Proof of the "Residue Theorem" for λ_a with $a \in N$: Suppose that</u> the assumptions of the Residue Theorem are made (page 35). As a partition of unity shows, it suffices to construct an open neighborhood $B(z_0)$ to every point $z_0 \in \overline{H}$, such that the "Residue Theorem" holds for $\mu\chi$ instead of χ, where μ is any C^∞-function on M with compact support in $B(z_0)$.

If $f(z_0) \neq a$, take an open, relative compact neighborhood $B(z_0)$ of z_0 with $a \notin f(\overline{B(z_0)})$. Let μ be any C^∞-function on M with compact support in $B(z_0)$. Then

$$\int_{H \cap f^{-1}(a)} \nu_f \psi\mu\chi = 0$$

and $d^\perp f^*(\lambda_a) \wedge \mu\chi$ is of class C^∞ on M. Hence, the Stoke's Theorem implies the formula of the Residue Theorem with $\mu\chi$ instead of χ.

Therefore, the case $f(z_0) = a$ has only to be considered where $z_0 \in \overline{H}$. Take $\alpha_j\colon W \to W_j'$ for $j = 1,\ldots,r$ and μ_j, ρ_j, κ_j such that condition 3) of Definition 5.1 holds. Because $z_0 \in \overline{H} \cap f^{-1}(a)$, the map f is adapted to $\mathfrak{A}_N$ at z_0 for a, which in this case of $s = n$ and $m - n = q$ means, that f is open in a neighborhood of z_0. Therefore an open, relative compact neighborhood $B(z_0)$ of z_0 exists such that $f(B(z_0)) \subseteq W$ and such that $f|B(z_0)$ is open. Let μ be a C^∞-function on M with compact support in $B(z_0)$. Define $H_0 = H \cap B(z_0)$ and $S_0 = S \cap B(z_0)$.

Take a test family $\{g_\rho\}_{\rho\in P}$. Take ρ_0 in $0 < \rho_0 < 1$ such that $w \in W_1'$ for all $w \in \mathbb{C}^n$ with $|w| \leqq \rho_0$. Define $\gamma_\rho: B(z_0) \to \mathbb{R}$ by $\gamma_\rho(z) = g_\rho(|\alpha_1(f(z))|)$ for $z \in B(z_0)$ and $0 < \rho < \rho_0$. Define $h_j = \alpha_1 \circ \alpha_j^{-1}$: $W_j' \to W_1'$. The map h_j is biholomorphic with $h_j(0) = 0$. Define $u_{\rho j}: W_j' \to \mathbb{R}$ by $u_{\rho j}(w) = g_\rho(|h_j(w)|)$. If $z \in B(z_0)$, then

$$u_{\rho j}(\alpha_j(f(z))) = g_\rho(|h_j(\alpha_j(f(z)))|)$$

$$= g_\rho(|\alpha_1(f(z))|) = \gamma_\rho(z).$$

Hence $u_{\rho j} \circ \alpha_j \circ f = \gamma_\rho$ does not depend on $j = 1,\ldots,r$. The intersection $K = \overline{H} \cap \operatorname{supp} \mu$ is a compact subset of $B(z_0)$. Define $\varphi_j = (\alpha_j^{-1})^* \rho_j$ on W_j'. Appendix I Proposition A I 15 implies

$$J_{j\rho} = \int_{H_0} d\gamma_\rho \wedge (\log|\alpha_j \circ f|)^{\kappa_j} \frac{1}{|\alpha_j \circ f|^{2n-2}} f^*(\rho_j) \wedge \psi\mu\chi$$

$$= \int_K d\gamma_\rho \wedge (\log|\alpha_j \circ f|)^{\kappa_j} \frac{1}{|\alpha_j \circ f|^{2n-2}} (\alpha_j \circ f)^*(\varphi_j) \wedge \psi\mu\chi$$

$$\to 0 \text{ for } \rho \to 0.$$

Let B_1 be an open, relative compact neighborhood of supp μ with $\overline{B}_1 \subset B(z_0)$. Define $H_1 = H \cap B_1$. Then $\overline{H}_s$ is a compact subset of $B(z_0)$. Let T_1 be the support of $\psi\mu\chi$ on $S_1 = \overline{H}_1 - H$. Because $T_1 \subseteq T$, the set $f^{-1}(a) \cap T_1$ is a set of measure zero on the analytic set $f^{-1}(a) = (\alpha_j \circ f)^{-1}(0)$ (respectively empty if $q = p$). Because $\alpha_j : W \to W'_j$ is biholomorphic $\nu_{\alpha_j \circ f}(z) = \nu_f(z)$ for $z \in B(z_0)$. Appendix I Theorem A I 18 implies

$$I_{j\rho} = \int_{H_0} d\gamma_\rho \wedge d^\perp \log \frac{1}{|\alpha_j \circ f|} \frac{1}{|\alpha_j \circ f|^{2n-2}} (\alpha_j \circ f)^*(\upsilon_{n-1}) \wedge \psi\mu\mu_j\chi$$

$$= \int_{H_1} d\gamma_\rho \wedge d^\perp \log \frac{1}{|\alpha_j \circ f|} \frac{1}{|\alpha_j \circ f|^{2n-2}} (\alpha_j \circ f^*)(\upsilon_{n-1}) \wedge \psi\mu\mu_j\chi$$

$$\to \frac{2\pi^n}{(n-1)!} \int_{H_1 \cap f^{-1}(a)} \nu_f \psi\mu\mu_j\chi$$

$$= \frac{2\pi^n}{(n-1)!} \int_{H \cap f^{-1}(a)} \nu_f \psi\mu\mu_j\chi$$

for $\rho \to 0$. Therefore

$$\int_{H_0} d\gamma_\rho \wedge d^\perp f^*(\lambda_a) \wedge \psi\mu\chi$$

$$= \frac{(n-1)!}{2\pi^n} \sum_{j=1}^{r} I_{j\rho} + \sum_{j=1}^{r} J_{j\rho}$$

$$\to \sum_{j=1}^{r} \int_{H \cap f^{-1}(a)} \nu_f \mu\mu_j(a)\psi\chi$$

.

$$= \int_{H \cap f^{-1}(a)} v_f \mu\psi\chi \qquad \text{for } \rho \to 0.$$

Stoke's Theorem implies

$$\int_{S_0} \gamma_\rho d^\perp f^*(\lambda_a) \wedge \psi\mu\chi + \int_{H_0} \gamma_\rho f^*(\omega_a) \wedge \psi\mu\chi$$

$$= \int_{H_0} \gamma_\rho d\psi \wedge d^\perp f^*(\lambda_a) \wedge \mu\chi - \int_{H_0} \gamma_\rho \psi d^\perp f^*(\lambda_a) \wedge d(\mu\chi)$$

$$+ \int_{H_0} d\gamma_\rho \wedge d^\perp f^*(\lambda_a) \wedge \psi\mu\chi.$$

By assumption $d^\perp f^*(\lambda_a) \wedge \psi\chi$ is integrable over S. Hence, also $d^\perp f^*(\lambda_a) \wedge \psi\mu\chi$ is integrable over S and S_0. Hence

$$\int_{S_0} \gamma_\rho d^\perp f^*(\lambda_a) \wedge \psi\mu\chi \to \int_{S_0} d^\perp f^*(\lambda_a) \wedge \psi\mu\chi$$

$\rho \to 0$, where S_0 may be replaced by S in the last integral.

According to Proposition 6.2 the integrals

$$\int_H d\psi \wedge d^\perp f^*(\lambda_a) \wedge \mu\chi = \int_{H_0} d\psi \wedge d^\perp f^*(\lambda_a) \wedge \mu\chi$$

$$\int_H \psi d^\perp f^*(\lambda_a) \wedge d(\mu\chi) = \int_{H_0} \psi \wedge d^\perp f^*(\lambda_a) \wedge d(\mu\chi)$$

exist. Trivially, the integral

$$\int_H \psi f^*(\omega_a) \wedge \mu\chi = \int_{H_0} \psi f^*(\omega_a) \wedge \mu\chi$$

exists. For $z \in \overline{H} - f^{-1}(a)$ (hence almost everywhere on H and S) $\gamma_\rho(z) \to 1$ for $\rho \to 0$, where $0 \leq \gamma_\rho \leq 1$. Therefore, $\rho \to 0$ implies:

$$\int_S \psi d^\perp f^*(\lambda_a) \wedge \mu\chi + \int_H \psi f^*(\omega_a) \wedge \mu\chi$$

$$= \int_H d\psi \wedge d^\perp f^*(\lambda_a) \wedge \mu\chi - \int_H \psi d^\perp f^*(\lambda_a) \wedge d(\mu\chi)$$

$$+ \int_{H \cap f^{-1}(a)} \nu_f \psi\mu\chi .$$

Hence, a partition of unity on $\overline{H}$ with the neighborhoods $B(z_0)$ proves the Residue Theorem.

<u>Proof of the "Singular Stoke's Theorem" for λ_a with $a \in N$:</u>

Suppose that the assumptions of the "Singular Stoke's Theorem" are made. As a partition of unity shows, it suffices to construct an open neighborhood $B(z_0)$ to every $z_0 \in \overline{H}$, such that the "Singular Stoke's Theorem" holds for $\mu\chi$ instead of χ, where μ is any C^∞-function on M with compact support in $B(z_0)$.

If $f(z_0) \neq a$, take an open, relative compact neighborhood $B(z_0)$ of z_0 with $a \notin f(\overline{B(z_0)})$. Let μ be any C^∞-function on M with compact support in $B(z_0)$. The usual Stoke's Theorem implies the formula of the Singular Stoke's Theorem with $\mu\chi$ instead of χ.

Therefore, only the case $f(z_0) = a$ has to be considered where $z_0 \in \overline{H}$. Take $\alpha\colon U \to U'$, V, κ and η such as in condition 2 of Definition 5.1. Without loss of generality $\alpha(a) = 0$ can be assumed. Because $z_0 \in f^{-1}(a) \cap \overline{H}$, the map f is adapted to $\mathfrak{A}_N$ at z_0 for a; hence f is open in a neighborhood of z_0. Take an open, relative compact neighborhood $B(z_0)$ of z_0 such that $f|B(z_0)$ is open, such that $f(\overline{B(z_0)}) \subset V$ and such that $S_0 = S \cap B(z_0)$ is either empty or contained in a component of S. Define $H_0 = H \cap B(z_0)$. Let μ be a C^∞-function on M with compact support in $B(z_0)$.

Take a test family[18)] $\{g_\rho\}_{\rho \in P}$. Take ρ_0 with $0 < \rho_0 < 1$ such that $w \in V' = \alpha(V)$ if $w \in \mathbb{C}^n$ with $|w| \leqq \rho_0$. Define $\gamma_\rho\colon B(z_0) \to \mathbb{R}$ by $\gamma_\rho(z) = g_\rho(|\alpha(f(z))|)$ for $z \in B(z_0)$ and $0 < \rho < \rho_0$. The intersection $K = \overline{H} \cap \operatorname{supp} \mu$ is a compact subset of $B(z_0)$. Define $\varphi = (\alpha^{-1})^*(\eta_a)$. Appendix I Proposition A I 15 (or Lemma A I 14) implies

$$\int_{H_0} d\gamma_\rho \wedge f^*(\lambda_a) \wedge d^\perp\psi \wedge \mu\chi$$

$$= \int_K d\gamma_\rho \wedge (\log |\alpha \circ f|)^\kappa \frac{1}{|\alpha \circ f|^{2n-2}} (\alpha \circ f)^*(\varphi) \wedge d^\perp\psi \wedge \mu\chi$$

$$\to 0 \text{ for } \rho \to 0$$

The usual Stoke's Theorem implies

$$\int_{S_0} \gamma_\rho f^*(\lambda_a) \wedge d^\perp\psi \wedge \mu\chi$$

$$= \int_{H_0} d\gamma_\rho \wedge f^*(\lambda_a) \wedge d^\perp\psi \wedge \mu\chi + \int_{H_0} \gamma_\rho f^*(\lambda_a) \wedge dd^\perp\psi \wedge \mu\chi$$

$$- \int_{H_0} \gamma_\rho f^*(\lambda_a) \wedge d^\perp\psi \wedge d(\mu\chi) + \int_{H_0} \gamma_\rho df^*(\lambda_a) \wedge d^\perp\psi \wedge \mu\chi$$

According to Proposition 6.1 the integrals

$$\int_H f^*(\lambda_a) \wedge dd^\perp\psi \wedge \mu\chi = \int_{H_0} f^*(\lambda_a) \wedge dd^\perp\psi \wedge \mu\chi$$

$$\int_H f^*(\lambda_a) \wedge d^\perp\psi \wedge d(\mu\chi) = \int_{H_0} f^*(\lambda_a) \wedge d^\perp\psi \wedge d(\mu\chi)$$

exist. According to Proposition 6.2. The integral

$$\int_H d\psi \wedge d^\perp f^*(\lambda_a) \wedge \mu\chi = \int_H df^*(\lambda_a) \wedge d^\perp\psi \wedge \mu\chi$$

$$= \int_{H_0} df^*(\lambda_a) \wedge d^\perp\psi \wedge \mu\chi$$

exists. Because $\gamma_\rho \to 1$ for $\rho \to 0$ with $0 \leqq \gamma_\rho \leqq 1$ the following limit exists

$$L = \lim_{\rho \to 0} \int_{S_0} \gamma_\rho f^*(\lambda_a) \wedge d^\perp\psi \wedge \mu\chi$$

$$= \int_H f^*(\lambda_a) \wedge dd^\perp\psi \wedge \mu\chi - \int_H f^*(\lambda_a) \wedge d^\perp\psi \wedge d(\mu\chi) +$$

$$+ \int_H df^*(\lambda_a) \wedge d^\perp\psi \wedge \mu\chi .$$

In the Singular Stoke's Theorem either assumption a) or b) are made concerning the boundary integral. Let $j: S \to M$ be the inclusion map. a) states that $j^*(f^*(\lambda_a) \wedge d^\perp\psi \wedge \chi)$ is assumed to be integrable over S. Then $j^*(f^*(\lambda_a) \wedge d^\perp \wedge \mu\chi$ is integrable over S_0 and S. Hence

$$L = \int_{S_0} f^*(\lambda_a) \wedge d^\perp\psi \wedge \mu\chi = \int_S f^*(\lambda_a) \wedge d^\perp\psi \wedge \mu\chi .$$

Assumption b) requires that the form $\varphi = j^*(f^*(\lambda_a) \wedge d^\perp\psi \wedge \chi)$ is either non-negative or non-positive on each component of S. Because S_0 is contained in at most one component of S (or $S_0 = \emptyset$), $\varepsilon = 1$ or $\varepsilon = -1$ exist such that $\varepsilon\varphi \geqq 0$ on S. Let C be any compact subset of $S_0 - f^{-1}(a)$. The minimum $\rho(C)$ of $|\alpha \circ f|$ on C is positive. If $z \in C$ and $0 < \rho < \rho(C)$, then $\gamma_\rho(z) = 1$. Hence

$$\int_C \varepsilon\, \mu\varphi \leqq \lim_{\rho \to 0} \int_{S_0} \gamma_\rho\, \varepsilon\, \mu\varphi = \varepsilon\, L < \infty.$$

Because $f^{-1}(a) \cap S_0$ is a set of measure zero on S_0, and because C is any compact subset of $S_0 - f^{-1}(a)$, this implies that $\varepsilon\mu\varphi$ and hence $\mu\varphi$ is integrable over S_0. Hence $L = \int_{S_0} \mu\varphi$ as in case a). In both cases

$$\int_S f^*(\lambda_a) \wedge d^\perp\psi \wedge \mu\chi$$

$$= \int_H f^*(\lambda_a) \wedge dd^\perp\psi \wedge \mu\chi - \int_H f^*(\lambda_a) \wedge d^\perp\psi \wedge d(\mu\chi)$$

$$+ \int_H df^*(\lambda_a) \wedge d^\perp\psi \wedge \mu\chi$$

Now, a partition of unity proves the Singular Stoke's Theorem, q.e.d.

§7. The construction of the proximity form.

Let N be a Kaehler manifold of dimension n. Let ω_1 be the fundamental form of bidegree (1,1) associated to the Kaehler metric. For $1 \leqq p \leqq n$ define

$$\omega_p = \omega_1 \wedge \cdots \wedge \omega_1 \qquad \text{(p-times)}.$$

Then ω_p is a positive form of bidegree (p,p) and class C^∞ on N with $d\omega_p = 0$ and $d^\perp\omega_p = 0$. Define $\omega_0 = 1$. Let T(N) be the cotangent bundle. The dual metric to the Kaehler metric on the tangent bundle, defines a metric along the fibers of the complexified cotangent bundle which in turn induces a metric along the fibers of $T^m(N)$ and $T^{p,q}(N)$, denoted by $(\,|\,)_x$ in the fiber over x. A bundle isomorphism

$$*: T^{p,q}(N) \to T^{n-p,n-q}(N)$$

exists uniquely, such that $\varphi \wedge *\psi = (\varphi|\psi)_x\omega_n(x)$ for all $\varphi \in T_x^{p,q}(N)$ and $\psi \in T_x^{p,q}(N)$ and $x \in N$. (If $p = q = 0$, define $(\varphi|\psi)_x = \varphi(x)\overline{\psi(x)}$.) Let $A^{p,q} = \Gamma^\infty(N, T^{p,q}(N))$ be the vectorspace of forms of bidegree (p,q) and of class C^∞ on N. Let $A^m = \bigoplus_{p+q=m} A^{p,q}$ be the vector space of forms of degree m and of class C^∞ on N. Define $A = \bigoplus_{m=0}^{2n} A^m$. Define[19)]

$$L: A \to A \qquad \text{by} \qquad L\varphi = \omega \wedge \varphi$$

Obviously $L: A^{p,q} \to A^{p+1,q+1}$. Then L commutes with d, $d^{\perp}$, ∂ and $\bar{\partial}$. Define

$$\Lambda = *^{-1}L* : A \to A$$

$$\delta = -*d* : A \to A$$

Then

$$\Lambda = (-1)^{p+q}*L*: A^{p,q} \to A^{p-1,q-1}$$

$$\delta : A^{p,q} \to A^{p-1,q-1} .$$

Define $\Delta = d\delta + \delta d$. Then $\Delta: A^{p,q} \to A^{p,q}$ is the __Laplace operator__ and commutes with $*$, d, $d^{\perp}$, δ, L and Λ. Moreover

$$\Delta = d^{\perp}\Lambda d - d\Lambda d^{\perp} + dd^{\perp}\Lambda - \Lambda d^{\perp}d$$

If φ has bidegree (n,n), then $d\varphi = 0$ and $d^{\perp}\varphi = 0$; hence

$$\Delta\varphi = dd^{\perp}(\Lambda\varphi).$$

Now, suppose that N is a compact Kaehler manifold. Define

$$(\varphi,\psi) = \int_N \varphi \wedge *\psi$$

for $\varphi \in A^m$ and $\psi \in A^m$. Then (,) is a hermitian product on A^m. Define the vector space of harmonic (p,q)-forms by

$H^{p,q} = \{\varphi \in A^{p,q} | \Delta\varphi = 0\}$ and $H^m = \bigoplus_{p+q=n} H^{p,q}$. A linear projection $h: A^{p,q} \to H^{p,q}$ with $h \circ h = h$ exists uniquely such that $(\varphi,\psi) = (h\varphi,\psi)$ for all $\varphi \in A^{p,q}$ and all $\psi \in H^{p,q}$. Then one and only one linear map $G: A^{p,q} \to A^{p,q}$ exists such that $G \circ h = h \circ G$ and

$$\mathrm{Id} - h = \Delta \circ G$$

Proposition 7.1. Let N be a compact Kaehler manifold of dimension n. Take $\varphi \in \Gamma^\infty(N\times N, \hat{T}^{p,q}(N))$. Then unique forms $G\varphi$ and $h\varphi$ in $\Gamma^\infty(N\times N, \hat{T}^{p,q}(N))$ exist such that $(G\varphi)_t = G(\varphi_t)$ and $(h\varphi)_t = h(\varphi_t)$. Moreover G and h are linear maps of $\Gamma^\infty(N\times N, \hat{T}^{p,q}(N))$ into itself.

Proof. Taking $t \in N$, fixed in the second factor N of $N\times N$, $G\varphi$ and $h\varphi$ are well-defined by the formulas indicated. It is known, that they are of class C^∞, for instance, Kodaira and Spencer [10] Theorem 7 page 65; q.e.d.

Obviously, also $\Lambda\varphi \in \Gamma^\infty(N\times N, \hat{T}^{p,q}(N))$ is well defined by $(\Lambda\,\varphi)_t = \Lambda(\varphi_t)$ and is of class C^∞.

Proposition 7.2. Let N be a connected, compact Kaehler manifold of dimension n. Take $\psi \in \Gamma^\infty(N\times N, \hat{T}^{n,n})$. Suppose that

$$\int_N \psi_t = 0$$

for each $t \in N$. Define $\varphi = -\Lambda G\psi$ in $\Gamma^\infty(N\times N, \hat{T}^{n-1,n-1})$. Then

$$\psi_t = d^{\perp}d\varphi_t \qquad \underline{\text{for each } t \in N.}$$

Proof. Take $g \in H^{n,n}$, then $*\overline{g} \in H^{0,0}$. Hence, $*\overline{g}$ is constant, because N is compact and connected. Therefore

$$(h\psi_t, g) = (\psi_t, g) = \int_N \psi_t \wedge *\overline{g} = *\overline{g} \int_N \psi_t = 0$$

Hence $h\psi_t = 0$. Therefore

$$\psi_t = \Delta G\psi_t = dd^{\perp}\Lambda G\psi_t = d^{\perp}d\varphi_t$$

q.e.d.

Lemma 7.3. Let N be a hermitian manifold of dimension n. Let A be a differentiable or complex manifold. Let ω_1 be the fundamental form of the hermitian metric on N. Define $\omega_p = \omega_1 \wedge \ldots \wedge \omega_1$. Let U be open and K be compact in N x A with $K \subseteq U$. For $t \in A$, define $U_t = U \cap (N \times \{t\})$ and $K_t = K \cap (N \times \{t\})$. Take $\chi \in \Gamma^0(U, \hat{T}^{p,p}(N))$ as a continuous section. Then a constant $C > 0$ exists such that $\chi_t + C\omega_t$ is positive on K_t for each $t \in N$ with $K_t \neq \emptyset$.

Proof. The projection K' of K into N is compact. Take $x_0 \in K'$. Then an open neighborhood V of x_0 on N and a biholomorphic map $\alpha = (z_1, \ldots, z_n) : V \to V'$ onto an open subset V' of $\mathbb{C}^n$ exists such that

$$\omega_1(x_0) = \frac{i}{2} \sum_{\mu=1}^{n} dz_\mu(x_0) \wedge d\overline{z}_\mu(x_0).$$

Then

$$\omega_p = (-1)^{\frac{p(p-1)}{2}} (\tfrac{1}{2})^p \sum_{\rho,\sigma\in T(p,n)} \omega_{\rho\sigma}\, dz_\rho \wedge d\bar{z}_\sigma$$

on V, where $\omega_{\rho\sigma}$ are C^∞-function on V with $\omega_{\rho\sigma}(x_0) = 0$ if $\rho \neq \sigma$ and $\omega_{\rho\sigma}(x_0) = p!$ if $\rho = \sigma$.

Define $\tilde{V} = (V\times A) \cap U$. Let $\tau: N \times A \to N$ be the projection. Then

$$\chi(x,t) = (-1)^{\frac{p(p-1)}{2}} (\tfrac{1}{2})^p \sum_{\rho,\sigma\in T(p,n)} \chi_{\rho\sigma}(x,t)\tau^*(dz_\rho \wedge d\bar{z}_\sigma)$$

on $\tilde{V}$, where the functions $\chi_{\rho\sigma}$ are continuous on $\tilde{V}$. A positive number C_0 exists such that

$$\sum_{\rho,\sigma\in T(p,n)} \chi_{\rho\sigma}(x_0,t)u_\rho\bar{u}_\sigma + p!C_0 \sum_{\rho\in T(p,n)} |u_\rho|^2 > 0$$

for all $t \in A$ with $(x_0,t) \in K$ and all vectors $u = \{u_\rho\}_{\rho\in T(p,n)}$ different from the zero vector. Hence an open neighborhood V_1 of x_0 exists such that

$$\sum_{\rho,\sigma\in T(p,n)} (\chi_{\rho\sigma}(x,t) + C_0\omega_{\rho\sigma}(x))u_\rho\bar{u}_\sigma > 0$$

for all $(x,t) \in (V_1\times A) \cap K$ and all non-zero vectors $u = \{u_\rho\}_{\rho\in T(p,n)}$.

Take $x_1 \in V_1 \cap K_{t_1}$. Then $(x_1,t_1) \in K$. Let L be a smooth complex submanifold of dimension p of V_1 with $x_1 \in L$. Let

$\beta = (w_1,\ldots,w_p) : W \to W'$ be a biholomorphic map of an open neighborhood of x_1 in W onto an open neighborhood W' of $\mathbb{C}^p$. Let $j: L \to V_1$ be the inclusion map. Then

$$j^*(dz_\rho) = u_\rho dw_1 \wedge \cdots \wedge dw_p$$

on W. Define $\upsilon_p = (\frac{1}{2})^p dw_1 \wedge d\overline{w}_1 \wedge \cdots \wedge dw_p \wedge d\overline{w}_p$. Then

$$j^*(\chi_{t_1} + C\omega)(x_1)$$

$$= \sum_{\rho,\sigma\in T(p,n)} (\chi_{\rho\sigma}(x_1,t_1) + C_0\omega_{\rho\sigma}(x_1))u_\rho(x_1)u_\sigma(x_1)\upsilon_p > 0$$

is positive. Therefore $\chi_t + C_0\omega_t > 0$ on $V_1 \cap K_t$. Finite by many neighborhoods $V_1^1,\ldots,V_1^q$ cover K'. For each V_1^μ a constant C_0^μ exists; let C be the largest among them. Then $\chi_t + C\omega_t > 0$ on K_t for each $t \in N$ with $K_t \neq \emptyset$; q.e.d.

Now the main existence theorem for proper proximity forms can be proved for a connected compact Kaehler manifold. The Kaehler metric can be always normalized such that the total volume $\int_N \omega_n = 1$.

Theorem 7.4. Let N be a connected, compact Kaehler manifold of dimension n. Let ω_1 be the fundamental form of the Kaehler metric and define $\omega_p = \omega_1 \wedge \cdots \wedge \omega_1$ (p-times). Suppose that the Kaehler metric is normalized such that $\int_N \omega_n = 1$.

<u>Then a singular potential $\lambda \in \Gamma^{\infty}(\hat{N}, \hat{T}^{n-1,n-1})$ for ω_n exists (more precisely for $\tau^*(\omega_n) \in \Gamma^{\infty}(N\times N, \hat{T}^n)$.</u>

<u>Proof</u>. According to Theorem 5.3 a singular potential $\tilde{\lambda}$ for some form $\chi \in \Gamma^{\infty}(N\times N, \hat{T}^{n,n})$ exists such that $\chi|G = 0$ in a neighborhood G of the diagonal F_N of N x N. According to Theorem 6.3 the Residue Theorem holds for $\tilde{\lambda}_t$ and χ_t and can be applied to the identity map f: $N \to N$ with $q = 0$, $\chi = 1$, $m = n$, $M = N = H$, $S = \emptyset$, $\psi = 1$, $T = \emptyset$. Hence

$$\int_N \chi_t = \nu_f^t(t) = 1.$$

Define $\psi = \tau^*(\omega_n) - \chi$, where $\tau: N \times N \to N$ is the projection onto the first factor. Then

$$\int_N \psi_t = \int_N \omega_n - \int_N \chi_t = 1 - 1 = 0.$$

According to Proposition 7.2 $\varphi \in \Gamma^{\infty}(N\times N, \hat{T}^{n-1,n-1})$ exists such that $\psi_t = d^{\perp}d\varphi_t$ for each $t \in N$. According to Lemma 7.3 a constant $C > 0$ exists such that $\varphi_t + C\omega_{n-1} > 0$ on N for every $t \in N$. Define

$$\lambda = \tilde{\lambda} + \varphi + C\tau^*(\omega_{n-1}) \in \Gamma^{\infty}(\hat{N}, T^{n-1,n-1})$$

Then $\lambda_t = \tilde{\lambda}_t + \psi_t + C\omega_{n-1} > 0$ on $N - \{t\}$ and

$$d^{\perp}d\lambda_t = d^{\perp}d\tilde{\lambda}_t + d^{\perp}d\varphi_t + 0 = \chi_t + \psi_t = \omega_n$$

on N - $\{t\}$. Clearly, conditions 2) and 3) of Definition 5.1 remain true for λ. Therefore, λ is a singular potential for $\tau^*(\omega_n)$ which means for ω_n; q.e.d.

Let N be a connected, compact Kaehler manifold of dimension n. Let ω_1 be the fundamental form and let ω_n be the volume element. Suppose that $\int_N \omega_n = 1$. Then any singular potential λ for ω_n is said to be a <u>proper proximity form</u>. Observe that for every $t \in N$, the form λ_t is proximity form of ω_n for t of the point family $\mathfrak{A}_N$ on N.

The First Main Theorem holds. Let M be a complex manifold of dimension m with $q = m - n \geq 0$. Let χ be a strictly non-negative form of class C^∞ and of bidegree (q,q) on M. Let $f: M \to N$ be a holomorphic map. If $K \subseteq M$, define

$$L(K) = \{a \in N \mid f \text{ adapted to } \mathfrak{A}_N \text{ at all } x \in K \text{ for } a\}$$

Then L(K) is the set of all $a \in N$, such that $f^{-1}(a) \cap K$ is either empty or $f|U$ is open for some neighborhood U of $f^{-1}(a) \cap K$. If K is compact, then L(K) is open.

Let $B = (G,\Gamma,g,\gamma,\psi)$ be a bump on M. Then

$$A_f(G) = \int_G f^*(\omega_n) \wedge \chi \qquad \text{(spherical image)}$$

$$T_f(G) = \int_G \psi f^*(\omega_n) \wedge \chi \qquad \text{(characteristic)}$$

If $a \in L(\overline{G})$, then

$$n_f(G,a) = \int_{G \cap f^{-1}(a)} \nu_f \chi \qquad \text{(counting function)}$$

$$N_f(G,a) = \int_{G \cap f^{-1}(a)} \nu_f \psi \chi \qquad \text{(integrated counting function)}$$

$$m_f(\Gamma,a) = \int_\Gamma f^*(\lambda_a) \wedge d^\perp \psi \wedge \chi \qquad \text{(proximity function)}$$

$$m_f(\gamma,a) = \int_\gamma f^*(\lambda_a) \wedge d^\perp \psi \wedge \chi \qquad \text{(proximity remainder)}$$

$$D_f(G,a) = \int_{G-g} f^*(\lambda_a) \wedge dd^\perp \psi \wedge \chi \qquad \text{(deficit)}$$

are defined. The formula of the <u>First Main Theorem</u> holds:

$$T_f(G) = N_f(G,a) + m_f(\Gamma,a) - m_f(\gamma,a) - D_f(G,a)$$

Here $N_f(G,a)$ is continuous on the open set $L(\overline{G})$ according to [27] Theorem 3.9. By Proposition 6.1, also $D_f(G,a)$ is continuous on $L(\overline{G})$. Now, take $\mu\colon M \to \mathbb{R}$ as a C^∞ function with $0 \leq \mu \leq 1$ such that μ has compact support in G and such that $\mu | U = 1$ where U is a neighborhood of g. Then $\tilde{B} = (G,\Gamma,g,\gamma,\mu\psi)$ is again a bump on M. Apply the first Main Theorem to $\tilde{B}$ where $\tilde{m}_f(\Gamma,a) = 0$ and $m_f(\gamma,a) = \tilde{m}_f(\gamma,a)$. Hence

$$\tilde{T}_f(G) = \tilde{N}_f(G,a) - m_f(\gamma,a) - \tilde{D}_f(G,a)$$

Hence $m_f(\gamma,a)$ is continuous on $L(\overline{G})$. Again the First Main Theorem implies that $m_f(\Gamma,a)$ is continuous on $L(\overline{G})$. Hence the following proposition is proved:

Proposition 7.5. For a proper proximity form on N the functions $N_f(G,a)$, $m_f(\gamma,a)$, $m_f(\Gamma,a)$ and $D_f(G,a)$ are continuous functions of a on the open set $L(\overline{G})$.

For the proper proximity form λ on the connected, compact Kaehler manifold N of dimension n, the average proximity form

$$\hat{\lambda} = L_{\omega_n}(\varphi) = \int_{t\in N} \omega_n(t) \otimes \lambda(\cdot,t)$$

exists and is a non-negative, continuous form of bidegree (n-1,n-1) (Proposition 6.3). For M,f,B as before, define the average proximity function by

$$\mu_f(\Gamma) = \int_{\Gamma} f^*(\hat{\lambda}) \wedge d^{\perp}\psi \wedge \chi \geqq 0$$

the average proximity remainder by

$$\mu_f(\gamma) = \int_{\gamma} f^*(\hat{\lambda}) \wedge d^{\perp}\psi \wedge \chi \geqq 0$$

Both integrals have non-negative integrands (Lemma 3.2). Define the average deficit by

$$\Delta_f(G) = \int_{G-g} f^*(\hat{\lambda}) \wedge dd^{\perp}\psi \wedge \chi$$

Proposition 7.6. Let N be a connected, compact Kaehler manifold of dimension n. Let λ be a proper proximity form on N. Let M be a complex manifold of dimension m with $q = m - n \geqq 0$. Let χ be a strictly non-negative form of bidegree (q,q) on M with $d\chi = 0$. Let $f: M \to N$ be a holomorphic map which is almost adapted to the point family $\mathfrak{A}_N$ of N. (i.e., which is open on some non-empty open subset of each component of M.) Then

a) $$A_f(G) = \int_{t\in N} n_f(G,t)\omega_n(t)$$

b) $$T_f(G) = \int_{t\in N} N_f(G,t)\omega_n(t)$$

c) $$\mu_f(\Gamma) = \int_{t\in N} m_f(\Gamma,t)\omega_n(t)$$

d) $$\mu_f(\gamma) = \int_{t\in N} m_f(\gamma,t)\omega_n(t)$$

e) $$\Delta_f(G) = \int_{t\in N} D_f(G,t)\omega_n(t)$$

f) $$\Delta_f(G) = \mu_f(\Gamma) - \mu_f(\gamma).$$

Proof. a) and b) are true by Theorem 4.2. Moreover,

$$\infty > \mu_f(\Gamma) = \int_\Gamma f^*\Big(\int_{t\in N}\omega_n(t) \otimes \lambda_t\Big) \wedge d^\perp\psi \wedge \chi$$

$$= \int_\Gamma \int_{t\in N}\omega_n(t) \otimes f^*(\lambda_t) \wedge d^\perp\psi \wedge \chi$$

$$= \int_{t\in N} \int_\Gamma \omega_n(t) \otimes f^*(\lambda_t) \wedge d^\perp\psi \wedge \chi$$

$$= \int_{t\in N} \left(\int_{\Gamma} f^*(\lambda_t) \wedge d^{\perp}\psi \wedge \chi \right) \omega_n(t)$$

$$= \int_{t\in N} m_f(\Gamma,t)\omega_n(t)$$

because all integrands are non-negative. This proves c). Obviously, d) is proved the same way.

If u is a continuous form of bidegree (1,1) on M, then the integral

$$D_f(G,a,u) = \int_{G-g} f^*(\lambda_a) \wedge u \wedge \chi$$

exists for every $a \in L(\overline{G})$ by Proposition 6.1. It is even a continuous function of a on $L(\overline{G})$. Observe that $N - L(\overline{G})$ has measure zero. Define

$$\Delta_f(G,u) = \int_{G-g} f^*(\hat{\lambda}) \wedge u \wedge \chi.$$

If $u \geqq 0$, then $f^*(\hat{\lambda}) \wedge u \wedge \chi \geqq 0$ and $f^*(\lambda_t) \wedge u \wedge \chi \geqq 0$. Hence

$$\Delta_f(G,u) = \int_{G-g} f^*\left(\int_{t\in N} \omega_n(t) \otimes \lambda_t \right) \wedge u \wedge \chi$$

$$= \int_{G-g} \int_{t\in N} (\omega_n(t) \otimes f^*(\lambda_t)) \wedge u \wedge \chi$$

$$= \int_{t\in N} \left(\int_{G-g} f^*(\lambda_t) \wedge u \wedge \chi \right) \omega_n(t)$$

$$= \int_{t\in N} D_f(G,t,u)\omega_n(t).$$

By definition $dd^{\perp}\psi$ is continuous on $\overline{G} - g$. Because M can be made into a hermitian manifold, Lemma 7.3 provides a positive, continuous form u of bidegree (1,1) on M such that $u + dd^{\perp}\psi = v$ is positive on $\overline{G}$. Then v can be continued to a non-negative, continuous form on M. Then

$$D_f(G,t,v) = D_f(G,t,u) - D_f(G,t)$$

$$\Delta_p(G,v) - \Delta_f(G,u) = \Delta_f(G).$$

Hence

$$\Delta_f(G) = \int_{t\in N} D_f(G,t,v)\omega_n(t) - \int_{t\in N} D_f(G,t,u)\omega_n(t)$$

$$= \int_{t\in N} D_f(G,t)\omega_n(t)$$

which proves e). Now, a) - e) and the First Main Theorem imply f), q.e.d.

Observe, if $\hat{\lambda}$ would be of class C^1 and if $d\hat{\lambda} = 0$ then f) would follow as a simple application of the Stoke's Formula. However, it is unknown if $\hat{\lambda}$ is of class C^1 or if this would be the case, if $d\hat{\lambda} = 0$. In the case of the Levine form (see [30] Theorem 2.11 and [28] Proposition 5.7) this is the case. If N is homogeneous, then

Hirschfelder [7] constructs an invariant proximity form λ. Then $\hat{\lambda}$ is also invariant, hence of class C^{∞}. If in addition N is symmetric, then $d\hat{\lambda} = 0$. However, these are special cases.

Now, the case of a general family shall be considered. For this, the following General Assumptions shall be made.

(A1) Let A be a connected, compact Kaehler manifold of dimension k. Let ω_1 be the fundamental form of the Kaehler metric. For $1 \leqq p \leqq k$ define $\omega_p = \omega_1 \wedge \cdots \wedge \omega_1$ (p-times); define $\omega_0 = 1$. Suppose that $\int_A \omega_k = 1$.

(A2) Let λ be a proper proximity form to ω_k on A for the point family $\mathfrak{A}_A$ on A. Define $\hat{\lambda} = L_{\omega_k}(\lambda)$ as the integral average.

(A3) Let $N \xleftarrow{\tau} F \xrightarrow{\pi} A$ be the defining triplet of an admissible $\mathfrak{A}$ of codimension $s > 0$ on the n-dimensional complex manifold N. Define $S_a = \tau(\pi^{-1}(a))$. Then $p = n - s$ is the dimension of S_a.

If these assumptions are made, then F and N are also compact. Moreover, the forms ω_k, λ_t, and $\hat{\lambda}$ can be lifted to F and integrated over the fibers of τ:

(A4) Define

$$\Omega = \tau_* \pi^*(\omega_k) \qquad \text{on } N$$

$$\Lambda_a = \tau_* \pi^*(\lambda_a) \qquad \text{on } N - S_a$$

$$\hat{\Lambda} = \tau_* \pi^*(\hat{\lambda}) \qquad \text{on } N$$

and call $\Lambda = \{\Lambda_a\}_{a \in A}$ a proper proximity form to Ω for the family $\mathfrak{A}$.

Observe, that Ω on N and Λ_a on $N - S_a$ are of class C^∞, where upon $\hat{\Lambda}$ is continuous on N. The dimension of F is $k + p$. Hence, the fiber dimension of τ is $k + p - n = p - s$. Hence Ω has bidegree (s,s). The forms Λ_a and $\hat{\Lambda}$ have bidegree $(s-1,s-1)$. Observe that Λ_a is a weak proximity form for Ω and for each $a \in A$.

(A5) Let M be a non-compact, complex manifold of dimension $m \geq s$. Define $q = m - n$. Then $n - p = s = m - q$.

(A6) Let χ be a strictly non-negative form of class C^∞ and of bidegree (q,q) on M with $d\chi = 0$. For $q = 0$ assume $\chi = 1$.

(A7) Let $f: M \to N$ be a holomorphic map. If $K \subseteq M$, define $L(K) = L(K, \mathfrak{A}, f)$ as the set of all $a \in A$ such that f is adapted to a for $\mathfrak{A}$ at all $x \in K$.

Always, $A - L(K)$ has measure zero. If K is compact then $L(K)$ is compact.

(A8) Suppose that f is almost adapted to $\mathfrak{A}$.

(A9) Define $f^*(F)$, $\tilde{f}$, $\hat{f}$ and σ as in §1 to obtain the diagram

$$\begin{array}{ccccc} \hat{f}: f^*(F) & \xrightarrow{\ \tilde{f}\ } & F & \xrightarrow{\ \pi\ } & A \\ \downarrow \sigma & 0 & \downarrow \tau & & \\ M & \xrightarrow{\ f\ } & N & & \end{array}$$

(A10) Let $B = (G,\Gamma,g,\gamma,\psi)$ be a bump on M.

If these assumptions are made, then the following value distribution functions are defined

$$A_f(G) = \int_G f^*(\Omega) \wedge \chi \qquad \text{(spherical image)}$$

$$T_f(G) = \int_G \psi f^*(\Omega) \wedge \chi \qquad \text{(characteristic)}$$

If $a \in L(\overline{G})$, then

$$n_f(G,a) = \int_{G\cap f^{-1}(S_a)} \nu_f^a \chi \qquad \text{(counting function)}$$

$$N_f(G,a) = \int_{G\cap f^{-1}(S_a)} \nu_f^a \psi \chi \qquad \text{(integrated counting function)}$$

$$m_f(\Gamma,a) = \int_\Gamma f^*(\Lambda_a) \wedge d^\perp\psi \wedge \chi \qquad \text{(proximity function)}$$

$$m_f(\gamma,a) = \int_\gamma f^*(\Lambda_a) \wedge d^\perp\psi \wedge \chi \qquad \text{(proximity remainder)}$$

$$D_f(G,a) = \int_{G-g} f^*(\Lambda_a) \wedge dd^\perp\psi \wedge \chi \qquad \text{(deficit)}$$

Because Λ_a is a weak proximity function for Ω, the First Main Theorem holds

$$T_f(G) = N_f(G,a) + m_f(\Gamma,a) - m_f(\gamma,a) - D_f(G,a)$$

for every $a \in L(\overline{G})$.

Define the average proximity function by

$$\mu_f(\Gamma) = \int_\Gamma f^*(\hat{\Lambda}) \wedge d^\perp\psi \wedge \chi$$

the <u>average proximity remainder</u> by

$$\mu_f(\gamma) = \int_\gamma f^*(\hat{\Lambda}) \wedge d^\perp\psi \wedge \chi.$$

Both integrals have non-negative integrands (Lemma 3.2). Define the <u>average deficit</u> by

$$\Delta_f(G) = \int_{G-g} f^*(\hat{\Lambda}) \wedge dd^\perp\psi \wedge \chi$$

A bump $B_\sigma = (G_\sigma, \Gamma_\sigma, g_\sigma, \gamma_\sigma; \psi\circ\sigma)$ is defined by setting $K_\sigma = \sigma^{-1}(K)$ for each subset K of M.

Now, build the value distribution functions for $\hat{f}, \mathfrak{A}_A, \lambda, \hat{\lambda}, \omega, f^*(F), \sigma^*(\chi), B_\sigma$, then it is shown exactly as in §4 page 46 - 49 that

$$A_f(G) = A_{\hat{f}}(G_\sigma) \qquad T_f(G) = T_{\hat{f}}(G_\sigma)$$

$$n_f(G,a) = n_{\hat{f}}(G_\sigma,a) \qquad N_f(G,a) = N_{\hat{f}}(G_0,a)$$

$$m_f(\Gamma,a) = m_{\hat{f}}(\Gamma_\sigma,a) \qquad m_f(\gamma,a) = m_{\hat{f}}(\gamma_\sigma,a)$$

$$D_f(G,a) = D_{\hat{f}}(G_\sigma,a) \qquad \Delta_f(G) = \Delta_{\hat{f}}(G_\sigma)$$

$$\mu_f(\Gamma) = \mu_{\hat{f}}(\Gamma_\sigma) \qquad \mu_f(\gamma) = \mu_{\hat{f}}(\gamma_\sigma)$$

for $a \in L(\overline{G}, \mathfrak{A}, f) = L(\overline{G}_\sigma, \mathfrak{A}_A, \hat{f})$. Therefore:

<u>Proposition 7.8.</u> <u>If the assumptions (A1) - (A7), (A9) and (A10) are made, then $N_f(G,a)$, $m_f(\Gamma,a)$, $m_f(\gamma,a)$ and $D_f(G,a)$ are continuous function on the open subset $L(\overline{G})$ of A, whose complement has measure zero. If also assumption (A8) is made, then</u>

$$T_f(G) = \int_A N_f(G,t)\omega_k(t)$$

$$A_f(G) = \int_A n_f(G,t)\omega_k(t)$$

$$\mu_f(\Gamma) = \int_A m_f(\Gamma,t)\omega_k(t)$$

$$\mu_f(\gamma) = \int_A m_f(\gamma,t)\omega_k(t)$$

$$\Delta_f(G) = \int_A D_f(G,t)\omega_k(t)$$

$$\Delta_f(G) = \mu_f(\Gamma) - \mu_f(\gamma)$$

Here, the question, if $\hat{\Lambda} = L_{\omega_k}(\Lambda)$, remains open.

For later applications, it will be important to know, when the characteristic is positive:

<u>Proposition 7.9.</u> <u>The assumptions (A1) - (A10) are made. Suppose that an open subset $U \neq \emptyset$ of M exists such that $\psi\chi|U > 0$. Then $A_f(G) > 0$ and $T_f(G) > 0$.</u>

Proof. Obviously, $U \subseteq G$. According to Proposition 2.7, $z_0 \in U$ and $a \in A$ exist such that $f(z_0) \in S_a$ and such that f is adopted to $\mathfrak{A}$ at z_0 for a. Let U_0 be an open neighborhood of z_0, such that $\overline{U}_0$ is compact and contained in U. Then $\psi \geqq \frac{1}{c} > 0$ on $\overline{U}_0$. Let $x_0 = (z_0, y_0) \in f^*(F)$ be the one and only point such that $\hat{f}(x_0) = \pi(y_0) = a$. An open neighborhood W of x_0 exists such that $\hat{f}(W)$ is open, such that $\sigma(W) \subseteq U_0$ and such that a biholomorphic map $\alpha: W \to W'$ onto an open neighborhood of $0 = \alpha(a)$ in $\mathbb{C}^{p+q}$ exists. Let W_0 be an open, relative compact neighborhood of a with $\overline{W}_0 \subset W$ such that $\alpha(W_0)$ is a ball with center 0 in W'. The image $V_0 = \hat{f}(W_0)$ is an open neighborhood of a.

For $t \in V_0$, define $\hat{F}_t = \hat{f}^{-1}(t) \cap W_0$ and $F_t = \sigma(\hat{F}_t)$, then $\sigma: \hat{F}_t \to F_t$ is a biholomorphic map into an open subset of the space $f^{-1}(S_t)$ with $F_t \subset U_0$. According to the continuity of the fiber integral [27] Theorem 3.9 the integral

$$L(t) = \int_{\hat{F}_t} v_{\hat{f}} \wedge \sigma^*(\chi) = \int_{F_t} v_f^t \chi \geqq 0$$

is a continuous function of t on V_0. Because $F_t \subseteq U_0$

$$L(t) \leqq n_f(G,t) \text{ and } L(t) \leqq cN_f(G,t)$$

for $t \in V_0$. Since $x_0 \in \hat{F}_a$, the set $\hat{F}_a$ is not empty and hence $L(a) > 0$. Therefore, $c_1 > 0$ and an open neighborhood V_1 of a with

$V_1 \subset V_0$ exist such that $L(t) \geqq c_1$ for $t \in V_1$. Because $V_1 \neq \emptyset$,

$$c_2 = \int_{V_1} \omega_k > 0$$

Then

$$A_f(G) \geqq \int_{V_1} n_f(G,t)\omega_k \geqq \int_{V_1} L(t)\omega_k \geqq c_1 c_2 > 0$$

$$T_f(G) \geqq \int_{V_1} N_f(G,t)\omega_k \geqq c\int_{V_1} L(t)\omega_k \geqq c c_1 c_2 > 0 .$$

Observe, that in order to prove $A_f(H) > 0$ only, it suffices to assume that $\chi|U > 0$ for some open subset $U \neq \emptyset$ of $H \subseteq M$.

§8 Level bumps.

In order to be able to obtain results on equidistribution, it is necessary to exhaust M by bumps. This can be done by the bumps associated to the level sets of an exhaustion function. Also, these level sets were used by Weyl [32], in [23] and [24] in a slightly different fashion. Here a situation which applies to both cases shall be studied. The value distribution functions shall be investigated in their dependency of the parameters.

Let M be a complex manifold M. Let $h: M \to \mathbb{R}$ be a non-negative, continuous function. For $r \geqq 0$, define

$$G_r = \{z \in M | h(z) < r\}$$

$$G'_r = \{z \in M | h(z) \leqq r\}$$

$$\Gamma_r = G'_r - G_r .$$

Then G_r is open; G'_r and Γ_r are closed with $G'_r \supseteq \overline{G}_r$. If $0 \leqq s < r$, then $G_s \subseteq G'_s \subseteq G_r$ and

$$G_r = \bigcup_{s<t<r} G_t \qquad G'_s = \bigcap_{s<t<r} G_t$$

Now, some additional assumptions will be made: At first, assume $r_0 \geqq 0$ and an open, subset $g \neq \emptyset$ are given, such that

a) The closure $\overline{g}$ is compact and $\gamma = \overline{g} - g$ is a boundary manifold of g.

b) It is $\overline{G}_{r_0} \subseteq g \subseteq \overline{g} \subseteq G'_{r_0}$.

Second, an interval I on the non-negative real axis is supposed to be given with $r_0 = \min I$ such that $\overline{G}_r$ is compact and such that h is of class C^∞ on $\overline{G}_r - g$ for all $r \in I$. Suppose that $\sup I = r_1 > r_0$.

Observe that $r_0 \in I$, but $r_1 \leqq \infty$ may be in I or not. For $r \in I$, define $\psi_r: M \to \mathbb{R}$ by

$$\psi_r(z) = \begin{cases} 0 & \text{if } z \in M - G_r, \text{ i.e., } h(z) \geqq r \\ r - h(z) & \text{if } z \in G_r - G_{r_0}, \text{ i.e., } r_0 \leqq h(z) \in r \\ r - r_0 & \text{if } z \in G_{r_0}, \text{ i.e., } h(z) < r_0 \end{cases}$$

On $G_r - \overline{g}$

$$d\psi_r = -dh \qquad d^\perp\psi_r = -d^\perp h \qquad dd^\perp\psi_r = d^\perp dh$$

does not depend on r. Define

$$I_h = \{r \in I \mid dh \neq 0 \text{ on } \Gamma_r\}$$

$$I_h^\circ = \{r \in I \mid \Gamma_r \text{ is a set of measure zero on } M\}.$$

Then $I_h \subseteq I_h^\circ$ and $I - I_h$ is a set of measure zero on $\mathbb{R}$. If $r \in I_h$, then $B_r = \{G_r, \Gamma_r, g, \gamma, \psi_r\}$ is a bump, and $\mathfrak{B}_h = \mathfrak{B}_h(I) = \{B_r\}_{r \in I_h}$ is a <u>family of bumps defined by h.</u> Two examples for the described situation shall be given

<u>Example 8.1 (Exhaustion function).</u> A non-negative function h of class C^∞ on M such that the map $h: M \to \mathbb{R}$ is proper, is said to be an exhaustion function. It always exists. Take one. Take $r_0 > 0$ such that $g = G_{r_0} \neq \emptyset$ and such that $dh \neq 0$ on $\gamma = \Gamma_{r_0}$. Then $\gamma = \overline{g} - g$ is a boundary manifold of g. Because h is proper, $\overline{G}_r$ is compact for each $r > 0$. Observe, $\sup h(M) = \infty$. Hence, I can be taken as $I = \{r \in \mathbb{R} | r \geqq r_0\}$. Obviously, the assumptions are satisfied with $r_1 = \infty$ and $G'_{r_0} = \overline{G}_{r_0} = \overline{g}$.

<u>Example 8.2.</u> Let $B = (G, \Gamma, g, \gamma, \psi)$. Define $h = R - \psi$ and $I = \{r \in R | 0 \leqq r \leqq R\}$. Obviously the previous assumptions are satisfied with $r_0 = 0$ and $r_1 = R$. Moreover,

$$G_r = \{z \in M | R - r > \psi(z)\}$$

and $g \subseteq G_r \subseteq G_R \subseteq G$ if $r \in I$ and $r > 0$. Moreover, $\psi_r(z) = r - R + \psi(z)$ if $z \in G_r$.

Now, return to the general case. The assumptions (A1) - (A7) and (A9) are made. Define $L(r) = L(\overline{G}_r)$ for each $r \geqq 0$. Then $L(r) \subseteq L(r')$ if $0 \leqq r' < r$. Define $L = L(M)$. For $r \in I_h$, abbreviate $A_f(r) = A_f(G_r)$, $T_f(r) = T_f(G_r)$, $m_f(\Gamma_r, a) = m_f(r, a)$, etc., replacing G_r or Γ_r by r. If $g = G_{r_0}$, write also $m_f(\gamma, a) = m_f(r_0, a)$, and $\mu_f(\gamma) = \mu_f(r_0)$. Observe, that $A_f(r)$, $T_f(r)$, and $\Delta_f(r)$ are defined for all $r \in I$ and this is also true for $n_f(r,a)$, $N_f(r,a)$ and $D_f(r,a)$ if $a \in L(r)$.

Lemma 8.3. Take $r \in I$ with $r > r_0$. Let U be a pure u-dimensional analytic subset of G_r. Let φ be a form of bidegree (u,u) on G_r which is integrable over U. Then

$$\int_U \psi_r \varphi = \int_{r_0}^{r} \left(\int_{U \cap G_t} \varphi \right) dt .$$

Proof. At first assume, that φ is non-negative at all simple points of U. Define $\rho(z,t) = 1$ if $z \in G_t$, and define $\rho(z,t) = 0$ if $z \in M - G_t$. Then $\rho(z,t) = 1$ if $t > h(z)$ and $\rho(z,t) = 0$ if $t \leqq h(z)$. Hence

$$\int_{r_0}^{r} \rho(z,t)\,dt = \left\{ \begin{array}{ll} 0 & \text{if } h(z) \geqq r \\ r - h(z) & \text{if } r_0 \leqq h(z) < r \\ r - r_0 & \text{if } h(z) < r_0 \end{array} \right\} = \psi_r(z)$$

for $z \in M$. Therefore,

$$\int_U \psi_r \varphi = \int_{z \in U} \int_{r_0}^{r} \rho(z,t)\ dt\ \varphi$$

$$= \int_{r_0}^{r} \left(\int_{z \in U} \rho(z,t) \varphi \right) dt$$

$$= \int_{r_0}^{r} \int_{U \cap G_t} \varphi\ dt .$$

In the general case, define $\mu^+(z) = 1$ (respectively $\mu^-(z) = 1$) if φ is non-negative (respectively negative) at the simple point z of U as form on U. At all other points of M define $\mu^+(z) = \mu^-(z) = 0$.

Then $\varphi = \mu^-\varphi + \mu^+\varphi$ at all simple points of U. Hence the formula holds for $\mu^-\varphi$ and $\mu^+\varphi$ respectively. Addition proves the Lemma, q.e.d.

Proposition 8.4. Under the assumptions made, the following statements hold:

1) A_f is an increasing function on I.

2) A_f, Δ_f and μ_f are continuous from the left at every $r \in I$ with $r > r_0$. Moreover, there are continuous at every $r \in I_h^0$ (if $r = r_0 \in I_h^0$, then only from the right).

3) T_f is increasing and continuous on I. At every $r \in I$ with $r > r_0$, T_f has a left sided derivative D^-T_f. At every $r \in I_h^0$ with $r_0 < r < r_1$, T_f is once differentiable. Moreover

$$T_f(r) = \int_{r_0}^{r} A_f(t)dt \quad \text{if } r \in I$$

$$(D^-T_f)(r) = A_f(r) \quad \text{if } r_0 < r \in I_h^0.$$

Proof. Because the integrand of A_f if non-negative, and because G_t is increasing in t, A_f is increasing. Because $G_r = \bigcup_{r_0<t<r} G_t$, because $dd^\perp\psi_r = d^\perp dh$ on $G_r - g$, the functions A_f, Δ_f and $\mu_f(r) = \Delta_f(r) + \mu_f(\gamma)$ are continuous from the left at every $r \in I$ with $r > r_0$. Because $G_r \cup \Gamma_r = \bigcap_{r<t<r_1} G_t$ and because Γ_r has measure zero for

$r \in I_h^0$, these functions are continuous from the right at every $r \in I_h^0$ with $r < r_1$. This proves 1) and 2). The integral representation in 3) follows from Lemma 8.3 with $U = G_r$ and $\varphi = f^*(\Omega) \wedge \chi$ for for $r > r_0$. For $r = r_0$, it is trivial. The integral representation and 1) and 2) imply the rest of 3), q.e.d.

Let $L^0(r)$ be the set of all $a \in L(r)$ such that $f^{-1}(S_a) \cap \Gamma_r$ has measure zero on $f^{-1}(S_a)$ if $q > 0$, respectively is empty, if $q = 0$.

<u>Proposition 8.5.</u> <u>Under the assumptions made the following statements hold:</u>

1. <u>If $r \in I$ and $a \in L_r$, then</u>

$$N_f(r,a) = \int_{r_0}^{r} n_f(t,a)\,dt.$$

2. <u>If r and t are in I with $t < r$ and if $a \in L(r)$, then</u>

$$0 \leqq n_f(t,a) \leqq n_f(r,a)$$

$$0 \leqq N_f(t,a) \leqq N_f(r,a).$$

3. <u>If $r \in I$ with $r > r_0$, if $a \in L(r)$ is fixed, then $n_f(\cdot,a)$, and $N_f(\cdot,a)$ are continuous from the left at r.</u>

4. If $r \in I$ with $r < r' \leqq r_1$, if $a \in L^0(r) \cap L(r')$, then $N_f(\cdot,a)$ and $n_f(\cdot,a)$ are continuous from the right (hence continuous if $r > r_0$) at r.

5. If t and r belong to I with $r_0 < t < r$ and if $a \in L(r)$, then $N_f(\cdot,a)$ is continuous at t.

6. If $r \in I$ with $r > r_0$ and if $a \in L(r)$, then the left sided derivative D^- of $N_f(\cdot,a)$ exists at r with $(D^-N_f)(r,a) = n_f(r,a)$.

7. If t and r belong to I with $r_0 < t < r$, and if $a \in L(r) \cap L^0(t)$, then $N_f(\cdot,a)$ is once differentiable at t with $N_f'(t,a) = n_f(t,a)$.

8. If $r \in I$ is fixed, then $N_f(r,\cdot)$ is continuous on $L(r)$.

9. If $r \in I$ is fixed, then $n_f(r,\cdot)$ is continuous at every $a \in L^0(r)$ as function on $L(r)$.

10. If t and r belong to I with $r_0 \leqq t < r$ and if $a \in L(r) \cap L^0(t)$, then n_f is continuous at $(t,a) \in I \times L(r)$.

11. If t and r belong to I with $r_0 \leqq t < r$ and if $a \in L(r)$, then N_f is continuous at $(t,a) \in I \times L(r)$.

Proof. Lemma 8.3 with $U = f^{-1}(S_a) \cap G_r$ and $\varphi = \nu_f^a\chi$ implies 1) for $r > r_0$. For $r = r_0$, it is trivial. Because $\nu_f^a\chi \geqq 0$, 2) follows. Because

$$G_r = \bigcup_{r_0<t<r} G_t\,, \quad G_r \cup \Gamma_r = \bigcap_{r<t<r_1} G_t$$

and because G_t is increasing in t, 3) and 4) follows. 5) follow from 1). 6) follows from 1) and 3). 7) is implied by 1) and 4).

Define $F(a) = \hat{f}^{-1}(a) \cap \sigma^{-1}(G_r)$. Then

$$n_f(r,a) = \int_{F(a)} v_{\hat{f}}\, \sigma^*(\chi)$$

$$N_f(r,a) = \int_{F(a)} v_{\hat{f}}\, (\psi_r \circ \sigma)\sigma^*(\chi)$$

for $a \in L(r)$, where $\psi_r \circ \sigma = 0$ on $\sigma^{-1}(\Gamma_r)$ and where $\sigma^{-1}(\Gamma_r) \cap \hat{f}^{-1}(a)$ is a set of measure zero if $a \in L^0(r)$. Hence the continuity of the fiber integral [27] Theorem 3.8 implies 8) and 9).

Take t and r in I with $r_0 < t < r$. Take $a \in L(r) \cap L^0(t)$. By Sards theorem a sequence $\{\eta_p\}_{p\in\mathbb{N}}$ exists such that $r_0 < t - \eta_p < t < t + \eta_p < r$, such that $\eta_p \to 0$ for $p \to \infty$ and such that $a \in L^0(t-\eta_p) \cap L^0(t+\eta_p)$ for all $p \in \mathbb{N}$. Then

$$n_f(t-\eta_p,b) \leqq n_f(x,b) \leqq n_f(t+\eta_p,b)$$

for all x with $t_p - \eta_p \leqq x \leqq t + \eta_p$ and all $b \in L(r)$. Now, 9) implies

$$n_f(t-\eta_p,a) \leqq \underline{\lim}\, n_f(x,b) \leqq \overline{\lim}\, n_f(x,b) \leqq n_f(t+\eta_p,a)$$

where the lower and upper limits are taken for $(x,b) \to (t,a)$. 3) and 4) imply

$$n_f(t,a) \leq \underline{\lim}\, n_f(x,b) \leq \overline{\lim}\, n_f(x,b) \leq n_f(t,a).$$

Hence $n_f(x,b) \to n_f(t,a)$ for $(x,b) \to (t,a)$. This proves 10) for $t > r_0$. The proof for $t = r_0$ is similar. 11) is proven the same way and the proof works for every $a \in L(r)$ because of 5), q.e.d.

Proposition 8.6. Under the assumptions made the following statements hold:

1. If $r \in I$ with $r > r_0$, if $a \in L(r)$ is fixed, then $D_f(\cdot,a)$ is continuous from the left at r.

2. If $r \in I_h^0$ and $r' \in I$ with $r < r' \leq r_1$, if $a \in L(r')$, then $D_f(\cdot,a)$ is continuous from the right at r (hence continuous at r if $r > r_0$).

3. If $r \in I$ is fixed, then $D_f(r,\cdot)$ is continuous on $L(r)$.

4. If $t \in I_h^0$ and $r \in I$ with $r_0 \leq t < r$ and if $a \in L(r)$ then D_f is continuous at $(t,a) \in I \times L(r)$.

Proof. Let u be a form of bidegree (1,1) on M with locally bounded coefficients. Define

$$D_f(r,u,a) = \int_{\sigma^{-1}(G_r - g)} \hat{f}^*(\lambda_a) \wedge \sigma^*(u \wedge \chi)$$

for $a \in L(r)$ and $r \in I$. By Proposition 6.1, this integral exists and is a continuous function of a on $L(r)$ for each fixed $r \in I$.

Moreover, fiber integration implies

$$D_f(r,u,a) = \int_{G_r - g} \sigma_* \hat{f}^*(\lambda_a) \wedge u \wedge \chi$$

$$= \int_{G_r - g} f^* \tau_* \pi^*(\lambda_a) \wedge u \wedge \chi$$

$$= \int_{G_r - g} f^*(\Lambda_a) \wedge u \wedge \chi$$

because $\sigma_* \hat{f}^* = \sigma_* \tilde{f}^* \pi^* = f^* \tau_* \pi^*$. If $u_0 = d^\perp dh$ on $\overline{G}_r - g$ and $u_0 = 0$ on the complement, then $D_f(r,u_0,a) = D_f(r,a)$. Therefore, 3) is true. Because $f^*(\Lambda_a) \wedge u \wedge \chi$ is integrable for each $a \in L(r)$. Because $L(r) \subseteq L(t)$ if $t \leqq r$, and because

$$G_r = \bigcup_{r_0 < t < r} G_t \qquad G_r \cup \Gamma_r = \bigcap_{r < t < r_1} G_t$$

$D_f(\cdot,u,a)$ is continuous from the left at $r \in I$ with $r > r_0$ if $a \in L(r)$ is fixed and $D_f(\cdot,u,a)$ is continuous from the right at $r \in I_h^0$ if $a \in L(r')$ for some $r' \in L$ with $r < r'$. Hence 1) and 2) are proved. Take $t \in I_h^0$ and $r \in I$ with $r_0 \leqq t < r$. Take $a \in L(r)$ A sequence $\{\eta_p\}_{p \in \mathbb{N}}$ exists such that $\eta_p \to 0$ for $p \to \infty$ and such that $r_0 < t - \eta_p < t < t + \eta_p < r$ (respectively $t < t + \eta_p$ only if $r_0 = t$).

Suppose that $u \geqq 0$ on G_r, then $f^*(\Lambda_a) \wedge u \wedge \chi \geqq 0$ on G_r. Therefore, if $r_0 < t < r$

$$D_f(t-\eta_p,u,b) \leqq D_f(x,u,b) \leqq D_f(t+\eta_p,u,b)$$

for all x with $t_p - \eta_p \leqq x \leqq t_p + \eta_p$ and all $b \in L(r)$. Now, 3) implies that

$$D_f(t-\eta_p,u,a) \leqq \underline{\lim}\, D_f(x,u,b) \leqq \overline{\lim}\, D_f(x,u,b)$$

$$\leqq D_f(t+\eta_p,u,a)$$

where the lower and upper limits are taken for $(x,b) \rightarrow (t,a)$. 1) and 2) imply

$$D_f(t,u,a) \leqq \underline{\lim}\, D_f(x,u,b) \leqq \overline{\lim}\, D_f(x,u,b) \leqq D_f(t,u,a) .$$

Hence $D_f(\cdot,u,\cdot)$ is continuous at (t,a). If $t = r_0$, the proof proceeds similarly.

Because $d^\perp dh$ is continuous on $\overline{G}_r$, a continuous non-negative form u of bidegree (1,1) on M and a constant $c > 0$ exist such that $v = d^\perp dh + cu \geqq 0$ on $\overline{G}_r$. For instance, let u be the fundamental form of a hermitian metric on M and determine $c > 0$ by Lemma 7.3. Then $D_f(\cdot,u,\cdot)$ and $D_f(\cdot,v,\cdot)$ are continuous at (t,a). Then $D_f = D_f(\cdot,v,\cdot) - D_f(\cdot,u,\cdot)$ is also continuous at (t,a), q.e.d.

According to Proposition 7.8, $m_f(\gamma,a)$ and $m_f(r,a)$ are continuous on L(r) for each fixed $r \in I_h$. Because $m_f(\gamma,a)$ is independent of r, it is continuous on

$$\bigcup_{r_0<r<r_a} L(r) = \bigcup_{r_1>r\in I_h} L(r).$$

Therefore, the continuity properties of m_f as functions of r and a is readily obtained from the First Main Theorem. The details are left to the reader. Observe, that $m_f(r,a)$ is only defined if $r \in I_h$ and $a \in L(r)$.

The previous results were rather complicated to state because the assumptions are very general. They become simpler, if more restricted assumptions are made.

Proposition 8.7. If in addition to the previous assumptions, f is adopted to $\mathfrak{A}$ at every $x \in M$ for every $a \in A$ and if $I_h^0 = I$, then A_f, T_f, Δ_f, μ_f are continuous on I, and N_f, D_f are continuous on I x A. Moreover, m_f can be continued to a continuous function on I x A. Moreover, T_f is differentiable on I with $T'_f = A_f$.

For A_f, T_f, Δ_f, μ_f this is an immediate consequence of Proposition 8.4. The continuity of N_f, D_f on $(I-\{r_1\}) \times A$ follows immediately from Proposition 8.5 and Proposition 8.6 and the continuity at (r_1,a) if $r_1 \in I$ and $a \in A$ follows by the same method using a one sided approach $r_1 - \eta_p < r_1 = t$ only.

Proposition 8.8. Suppose that (A1) - (A10) hold. Take h and I as before. Suppose that $r_1 = \infty$. Suppose that $\chi > 0$ on some open, non empty subset of M. Then

$$\infty \geqq \underline{\lim}_{r \to \infty} \frac{T_f(r)}{r} \geqq \lim_{r \to \infty} A_f(r) > 0.$$

Especially, $T_f(r) \to \infty$ for $r \to \infty$.

<u>Proof</u>. Let $U \neq \emptyset$, be an open, relative compact subset of M such that $\chi | U > 0$. Take $s \in I$ such that $s > h(z)$ for all $z \in U$. If $r \geqq s$, then $U \subset G_r$ and $\psi_r > 0$ on U. By Proposition 7.9, $A_f(r) > 0$ and $T_f(r) > 0$ if $r \geqq s$. Moreover, if $r > t \geqq s$, then

$$T_f(r) = \int_{r_0}^{r} A_f(\tau) d\tau \geqq (r-t) A_f(t)$$

Hence

$$\lim_{r \to \infty} \frac{T_f(r)}{r} \geqq A_f(t) > 0$$

Because A_f is increasing, the Proposition is proved, q.e.d.

§9 Equidistribution.

Let M be a non-compact, complex manifold of dimension m. Let I be a <u>directed set</u>, that is, I is partially order, and for any two elements r_1, r_2 in I an element r_3 exists such that $r_3 \geqq r_1$ and $r_3 \geqq r_2$. Observe, that any function on I is a net, and the concept of a limit on I is defined.

A family $\mathfrak{B} = \{B_r\}_{r\in I}$ is said to <u>exhaust M</u>, if and only if for every $r \in I$ of the directed set I, a bump $B_r = (G_r, \Gamma_r, g, \gamma, \psi_r)$ is given where g and γ are independent of r for all $r \in I$. Moreover, it is required, that for every compact subset K of M, an element $r_K \in I$ exists such that $\psi_r(z) > 0$ if $z \in K$ and $r \geqq r_K$ (especially, $G_r \supset K$ for $r \geqq r_K$).

Assumptions (A1) - (A9) are made now. In addition assume:

(A11) <u>The form χ of (A6) is positive on some open non-empty subset of M.</u>

Hence, an open, relative compact subset U of M exists such that $\chi|U > 0$. If $r \geqq \tilde{r} = r_{\overline{U}}$, then $\psi_r > 0$ on U. Hence

$$A_f(G_r) > 0 \qquad \text{and} \qquad T_f(G_r) > 0 \quad \text{if} \quad r \geqq \tilde{r}.$$

Define

$$J_f = \{a \in A | f^{-1}(S_a) \neq \phi\}.$$

Then $J_f = \{a \in A | \hat{f}^{-1}(a) \neq \phi\} = \hat{f}(f^*(F))$ is measurable in A. Define

$$b_f = \int_{J_f} \omega_k$$

Obviously, $0 \leqq b_f \leqq 1$ by (A1) and

$$0 \leqq 1 - b_f = \int_{A-J_f} \omega_k \leqq 1$$

is the measure of the set of points $a \in A$ with $f^{-1}(S_a) = \emptyset$. If $B = (G,\Gamma,g,\gamma,\psi)$ is a bump on M, define

$$\Delta_f^0(G) = \int_{J_f} D_f(G,t)\omega_k(t).$$

The integral exists, because $D_f(G,t)$ is integrable over A by Proposition 7.8.

<u>Proposition 9.1.</u> <u>If the assumptions (A1) - (A11) are made, then</u>

$$0 \leqq (1-b_f)T_f(G) \leqq \Delta_f^0(G) + \mu_f(\gamma).$$

<u>Proof</u>. The definitions of J_f and N_f imply

$$\int_{J_f} N_f(G,t)\omega_k(t) = \int_A N_f(G,t)\omega_k(t) = T_f(G)$$

Moreover,

$$\int_{J_f} m_f(\gamma,t)\omega_k(t) \leqq \int_A m_f(\gamma,t)\omega_k(t) = \mu_f(\gamma).$$

The First Main Theorem implies

$$N_f(G,t) - T_f(G) \leqq D_f(G,t) + m_f(\gamma,t)$$

for $t \in L(\overline{G})$, i.e., almost everywhere on A. Hence the integration over J_f proves the proposition, q.e.d.

If $\mathcal{B} = \{B_r\}_{r\in I}$ is a family of bumps which exhausts M, and if (AI) - (A10) and (A11) are assumed, then $T_f(G_r) > 0$ for $r \geqq \tilde{r}$. Hence

$$\Delta_f^0(\mathcal{B}) = \limsup \frac{\Delta_f^0(G_r)}{T_f(G)}$$

$$\Delta_f(\mathcal{B}) = \limsup \frac{\Delta_f(G_r)}{T_f(G_r)}$$

$$\mu_f(\mathcal{B}) = \limsup \frac{\mu_f(\gamma)}{T_f(G_r)} \geqq 0$$

are defined. Proposition 9.1 implies:

<u>Theorem 9.2.</u> <u>Suppose that the assumptions (A1) - (A9) and (A11) are made. If the family $\mathcal{B}$ of bumps exhausts M, then</u>

$$0 \leqq 1 - b_f \leqq \Delta_f^0(\mathcal{L}) + \mu_f(\mathcal{L}).$$

<u>If $\mu_f(\mathcal{L}) = \Delta_f^0(\mathcal{L}) = 0$, then $f(M) \cap S_a \neq \emptyset$ for almost all $a \in A$.</u>

The explicit value of this theorem is small. It is more a guide for further investigations. In special cases, $\Delta_f^0(\mathcal{L})$ and $\mu_f(\mathcal{L})$ shall be estimated or computed:

<u>I Case:</u> <u>Divisor Case.</u>[20] <u>Here $s = m - q = n - p = 1$ is assumed. Hence S_a and $f^{-1}(S_a)$ - if not empty - are divisors. (A1)-(A11) are assumed. Further assume that $\chi > 0$ everywhere on M and that M is connected.</u>

Let $\alpha: U \to U'$ be a biholomorphic map of an open subset U of M onto an open subset U' of $\mathbb{C}^m$. Set $\alpha = (z_1, \ldots, z_m)$. On U

$$\chi = \left(\tfrac{i}{2}\right)^{m-1} \sum_{\substack{\mu,\nu=1 \\ \mu \neq \nu}}^{m} \chi_{\mu\nu} d\bar{z}_\mu \wedge dz_\nu \wedge \prod_{\substack{\rho=1 \\ \rho \neq \mu,\nu}} dz_\rho \wedge d\bar{z}_\rho$$

$$+ \left(\tfrac{i}{2}\right)^{m-1} \sum_{\mu=1}^{m} \chi_{\mu\mu} \prod_{\substack{\rho=1 \\ \rho \neq \mu}} dz_\rho \wedge d\bar{z}_\rho$$

Then $\chi > 0$ means that for every $z \in U$ and every $x = (x_1, \ldots, x_m) \neq 0$ in $\mathbb{C}^m$

$$\sum_{\mu,\nu=1}^{m} a_{\mu\nu}(z) x_\mu \bar{x}_\nu > 0.$$

Suppose that φ is of class C^2 on U. Then

$$d^{\perp}d\varphi \wedge \chi = 4\left(\sum_{\mu,\nu=1}^{m} \chi_{\mu\nu}\varphi_{z_\mu \bar{z}_\nu}\right)\upsilon_m$$

$$d^{\perp}\varphi \wedge d\varphi \wedge \chi = 4 \sum_{\mu,\nu=1}^{m} \chi_{\mu\nu}\varphi_{z_\mu}\varphi_{\bar{z}_\nu}\upsilon_m .$$

Hence $d^{\perp}d\varphi \wedge \chi = 0$ is an elliptic differential equation for φ. If φ is real, then

$$d^{\perp}\varphi \wedge d\varphi \wedge \chi \geqq 0$$

and $d^{\perp}\varphi \wedge d\varphi \wedge \chi = 0$ at z implies $d\varphi = 0$ at z.

Now, a family $\mathcal{B} = \mathcal{B}(g,\chi)$ of bumps for χ shall be constructed on M. Take any open, relative compact subset $g \neq \phi$ of M such that $\gamma = \bar{g} - g$ is a boundary manifold of g and such that no component of M - g is compact. Such a set g exists. Because M is not compact and connected, $\gamma \neq \phi$.

Let I be the set of all open, connected, relative compact subsets G of M with $G \supset \bar{g}$ such that $\Gamma = \bar{G} - G$ is a boundary manifold of G. Observe, that each component of $G - \bar{g}$ has boundary points on Γ and γ. The set I is directed and

$$M = \bigcup_{G \in I} G$$

For $G \in I$, the Dirichlet problem, $dd^{\perp}\varphi \wedge \chi = 0$ on $G - \bar{g}$, and $\varphi|\gamma = 1$ and $\varphi|\Gamma = 0$ has one and only one solution $\varphi = \varphi_G$ and this solution is of class C^∞ on $\bar{G} - g$.[21] Because each component of $G - \bar{g}$

borders on Γ and γ, φ_G is not constant on any open subset of $G - \overline{g}$, with $0 < \varphi_G < 1$ on $G - \overline{g}$. This is due to the maximum principle. Consequently, $d^{\perp}\varphi \wedge d\varphi \wedge \chi > 0$ on $G - \overline{g}$. The <u>capacity of (G,g,χ)</u> is defined by

$$0 < C(G) = \int_{G-\overline{g}} d^{\perp}\varphi_G \wedge d\varphi_G \wedge \chi .$$

Stoke's Theorem implies

$$C(G) = \int_{\gamma} \varphi_G d^{\perp}\varphi_G \wedge \chi = \int_{\gamma} d^{\perp}\varphi_G \wedge \chi = \int_{\Gamma} d^{\perp}\varphi_G \wedge \chi .$$

because $dd^{\perp}\varphi_G \wedge \chi = 0$ on $G - \overline{g}$ and because of the boundary conditions. Define $\varphi_G = 1$ on g and $\varphi_G = 0$ on $M - \overline{G}$.

<u>Lemma 9.3.</u>[22] <u>If G_1 and G_2 belong to I with $G_1 \subset G_2$, then $C(G_2) \leqq C(G_1)$ and $\varphi_{G_1} \leqq \varphi_{G_2}$ on M.</u>

<u>Proof</u>. Define $\xi = \varphi_{G_2} - \varphi_{G_1}$. Then $\xi \geqq 0$ on $M - (G_1 - \overline{g})$. Moreover, $\xi|\gamma = 0$ and $\xi|\Gamma_1 \geqq 0$ with $dd^{\perp}\xi \wedge \chi = 0$ on $G_1 - \overline{g}$. The maximum principle implies $\xi \geqq 0$ on $G_1 - \overline{g}$. Let $j\colon \gamma \to M$ be the inclusion. Lemma 3.2 implies $j^*(d^{\perp}\xi \wedge \chi) \leqq 0$. Hence

$$C(G_2) - C(G_1) = \int_{\gamma} j^*(d^{\perp}\xi \wedge \chi) \leqq 0.$$

q.e.d.

Therefore, the total capacity of M for g and χ can be defined by

$$0 \leq C(M) = \inf_{G\in I} C(G) = \lim_{G\in I} C(G) < \infty.$$

If $C(M) = 0$, then M is said to have zero capacity for χ and g.

For $G \in I$, define $R(G) = \frac{1}{C(G)}$. Then $R(G_1) \leq R(G_2)$ if $G_1 \in I$ and $G_2 \in I$ with $G_1 \subset G_2$. Define $\psi_G = R(G)\varphi_G$ on M. Then ψ_G is continuous on M, of class C^∞ on $\overline{G} - g$, with $\psi_G|g = R(G) > 0$ and $\psi_G(M-G) = 0$. Moreover, $dd^\perp\psi_G \wedge \chi = 0$ and $0 < \psi_G < R(G)$ on $G - \overline{g}$. Also

$$\int_\Gamma d^\perp\psi_G \wedge \chi = 1 = \int_\gamma d^\perp\psi_G \wedge \chi,$$

For $G \in I$, the collection $B_G = \{G,\Gamma,g,\gamma,\psi_G\}$ is a bump. The family $\mathfrak{L} = \mathfrak{L}(g,\chi) = \{B_G\}_{G\in I}$ of bumps for g and χ exhausts M.

If G_1 and G_2 belong to I, and if $G_1 \subset G_2$, then

$$\psi_{G_1} = R(G_1)\varphi_{G_1} \leq R(G_2)\varphi_{G_2} \leq \psi_{G_2}$$

$$T_f(G_1) = \int_M \psi_{G_1} f^*(\Omega) \wedge \chi \leq \int_M \psi_{G_2} f^*(\Omega) \wedge \chi = T_f(G_2).$$

Therefore T_f is an increasing function of G on I. Define

$$T_f(M) = \sup_{G\in I} R(G) = \lim_{G\in I} R(G) \leq \infty$$
$$R(M) = \sup_{G\in I} R(G) = \lim_{G\in I} R(G) \leq \infty$$

<u>Proposition 9.4.</u> <u>If $R(M) = \infty$, then $T_f(M) = \infty$.</u>

<u>Proof</u>. Take $G \in I$, then

$$T_f(G) \geqq \int_g R(G) f^*(\Omega) \wedge \chi = R(G) A_f(g).$$

According to the remark at the end of §7, $A_f(g) > 0$. Now, $R(M) = \sup_{G \in I} R(G) = \lim_{G \in I} R(G) = \infty$ implies $T_f(M) = \infty$, q.e.d.

If $G \in I$, then $dd^{\perp}\psi_G \wedge \chi = 0$ on $G - \overline{g}$. Hence

$$\Delta_f(G,a) = 0 = \Delta_f(G) = \Delta_f^0(G)$$

$$\Delta_f^0(\mathcal{L}) = 0.$$

Observe that $s = 1$. Hence $\hat{\Lambda}$ is a non-negative, continuous function on the compact manifold N. Let Λ^0 be its maximum.

<u>Theorem 9.5.</u> <u>Suppose that the assumptions (A1)-(A9) are made. Suppose that $\chi > 0$ on M. Suppose that M is connected. Let $g \neq \emptyset$ be an open, relative compact subset of M such that $\gamma = \overline{g} - g$ is a boundary manifold of g and such that no component of M - g is compact. Construct I and $\mathcal{L} = \mathcal{L}(g,\chi)$. Let Λ^0 be the maximum of $\hat{\Lambda}$ on N.</u>

<u>If $T_f(M) = \infty$, then $b_f = 1$, i.e., $f(M) \cap S_a \neq \emptyset$ for almost all $a \in A$. (Observe, $C(M) = 0$ implies $T_f(M) = \infty$).</u>

<u>If $T_f(M) < \infty$, then $T_f(M) > 0$, and the measure of all $a \in A$</u>

with $f(M) \cap S_a = \emptyset$ is estimated by

$$0 \leqq 1 - b_f \leqq \frac{\Lambda^0}{T_f(M)}.$$

Proof. If $G \in I$, then

$$\mu_f(\gamma) = \int_\gamma (\hat{\Lambda}\circ f) d^\perp \psi_G \wedge \chi \leqq \Lambda^0 \int_\gamma d^\perp \psi_G \wedge \chi = \Lambda^0 < \infty$$

Hence $\mu_f(\mathcal{E}) = 0$ if $T_f(M) = \infty$. Obviously, $T_f(M) > 0$. Hence $\mu_f(\mathcal{E}) \leqq \Lambda^0 (T_f(M))^{-1}$ if $T_f(M) < \infty$, q.e.d.

Of course, the estimate for $1 - b_f$ says nothing if $\Lambda^0 \geqq T_f(M)$. Observe, that Λ^0 depends on (A1)-(A4) only. Especially, Λ^0 does not depend on f. Hence, if f grows so strong to overcome the critical mass Λ^0, i.e., if $\Lambda^0 < T_f(M)$, a meaningful estimate of the measure of the set of all $a \in A$ with $f(M) \cap S_a = \emptyset$ is obtained. Therefore, Theorem 9.5 can be considered as a quite satisfactory result of equidistribution theory if s = 1. With [24] in mind, a better result can be hoped for by the establishment of a defect relation, a problem which seems to be most difficult.

II Case: pseudoconcave manifolds. (A1)-(A11) are assumed. Suppose, that M is connected. Let h be a non-negative function of class C^∞ on M such that $h: M \to R$ is proper and such that $d^\perp dh \leqq 0$ outside a compact subset of M. Then M is called pseudoconcave. h is an exhaustion function in the sense of Example 8.1. Construct $r_0 > 0$, $g = G_{r_0}$, $I = \{r | r \geqq r_0\}$ and $\mathcal{B}_h = \mathcal{B}_h(I)$ as there. Moreover, r_0, (i.e., g) can be taken so large, that $d^\perp dh \leqq 0$ on $M - g$. Adopt the notations of §8. Then

$$D_f(r,a) = \int_{G_r - g} f^*(\Lambda_a) \wedge d^\perp dh \wedge \chi \leqq 0$$

if $r > r_0$ and $a \in L(r)$. Because $A - L(r)$ has measure zero,

$$\Delta_f^0(r) = \int_{J_f} D_f(r,a)\omega_k(a) \leqq 0$$

if $r > r_0$, which implies $\Delta_f^0(\mathcal{B}_h) \leqq 0$. Moreover,

$$\mu_f(\gamma) = -\int_\gamma f^*(\hat{\Lambda}) \wedge d^\perp h \wedge \chi$$

is constant in r. By Proposition 8.8, $T_f(r) \to \infty$ for $r \to \infty$. Hence $\mu_f(\mathcal{B}_h) = 0$. Theorem 9.2 implies:

Theorem 9.6. Assume (A1)-(A11). Suppose that the complex manifold M is connected and pseudoconcave. Then $b_f = 1$, i.e., $f(M) \cap S_a \neq \emptyset$ for almost all $a \in A$.

Pseudoconcave manifolds as defined here, are not too far away from compact manifolds; therefore, Theorem 9.6 does not seem too surprising.

III Case: Pseudoconvex case. (A1)-(A11) are assumed. Suppose that M is connected. Let h be a non-negative function of class C^∞ on M such that $h: M \to \mathbb{R}$ is proper, and such that $d^\perp dh \geqq 0$ outside a compact subset of M. Then M is called pseudoconvex. h is an exhaustion function in the sense of Example 8.1. Construct $r_0 > 0$, $g = G_{r_0}$, I and $\mathcal{B}_h = \mathcal{B}_h(g)$ as there. Moreover, r_0, i.e., g, shall be taken so large that $d^\perp dh \geqq 0$ on M - g. Adopt the notations of §8. Then

$$D_f(r,a) = \int_{G_r - g} f^*(\Lambda_a) \wedge d^\perp dh \wedge \chi \geqq 0$$

if $r > r_0$ and $a \in L(r)$. Because A - L(r) has measure zero,

$$\Delta_f^0(r) = \int_{J_f} D_f(r,a)\omega_k(a) \leqq \int_A D_f(r,a)\omega_k(a) = \Delta_f(r)$$

for $r \geqq r_0$. Hence

$$\Delta_f^0(\mathcal{B}_h) \leqq \limsup_{r \to \infty} \frac{\Delta_f(r)}{T_f(r)} = \Delta_f(\mathcal{B}_h)$$

Again, $T_f(r) \to \infty$ for $r \to \infty$ and $\mu_f(\gamma)$ is constant in r. Hence $\mu_f(\mathcal{B}_h) = 0$. Theorem 9.2 implies

Theorem 9.7. Assume (A1)-(A11). Suppose that the complex manifold M is connected and pseudoconvex. Then

$$0 \leqq 1 - b_f \leqq \limsup_{r \to \infty} \frac{\Delta_f(r)}{T_f(r)}$$

Especially, if $\Delta_f(r) = (T_f(r))$ for $r \to \infty$, then $f(M) \cap S_a \neq \emptyset$ for almost all $a \in A$.

This case of pseudoconvex manifolds contains the Stein manifolds and move, therefore, the additional condition $\Delta_f(r) = (T_f(r))$ certainly is not superfluous, as the example of the unit disk shows.

IV Case: Levi manifolds. The couple (M,h) is said to be a Levi manifold, if M is a non-compact, connected complex manifold, and if $h\colon M \to \mathbb{R}$ is a proper map of class C^∞ with $h \geqq 0$ and $d^\perp dh \geqq 0$ on M and such that $d^\perp dh > 0$ on some open, non-empty subset of M. Of course, this case falls under III, but it will be somewhat different. Only, (A1)-(A5) and (A7)-(A9) are assumed. Again h is an exhaustion function in the sense of Example 8.1. Construct $r_0 > 0, g = G_{r_0}, I = \{r \mid r \geqq r_0\}$ and $\mathcal{B}_h = \mathcal{B}_h(g)$. Now, χ shall have to be constructed. If $q = 0$, define $\chi_0 = 1$ as usual. Define

$$\chi_j = d^\perp dh \wedge \dots \wedge d^\perp dh \qquad (j\text{-times})$$

for $j = 1,\dots,m$. Then χ_j is strictly non-negative, of class C^∞ and positive on some non empty open subset of M. Moreover, $d\chi_j = 0$ on M. Now, (A6) and (A11) are satisfied with $\chi = \chi_q$. Now, adopt the notation of §8. Of course, Theorem 9.7 holds. For $1 \leqq j \leqq s$ define

$$\Omega_j = \tau_* \pi^*(\omega_{k-s+j}).$$

Then Ω_j is a non-negative form of bidegree (j,j) and class C^∞ on M. For $j = s, \Omega_s = \Omega$. For $1 \leqq j \leqq s$, define <u>spherical image A_{jf} and the characteristic T_{jf} of codimension j</u> by

$$A_{jf}(r) = \int_{G_r} f^*(\Omega_j) \wedge \chi_{m-j}$$

$$T_{jf}(r) = \int_{G_r} \psi_r f^*(\Omega_j) \wedge \chi_{m-j} .$$

Clearly, A_{jf} is continuous from the left and increasing. Lemma 8.3 implies

$$T_{jf}(r) = \int_{r_0}^{r} A_{jf}(t)dt \qquad \text{if } r \in I$$

and $D^- T_{jf}(r) = A_{jf}(r)$. Because A_{jf} is continuous at every $r \in I_h^0, T'_{jf}(r) = A_{jf}(r)$ for $r \in I_h^0$. Clearly, T_{jf} is continuous and increasing on I. For $j = s$

$$A_{sf}(r) = A_f(r) \qquad T_{sf}(r) = T_f(r).$$

Observe, that for $0 \leqq j < s$, $A_{jf}(r)$ and $T_{jf}(r)$ are not defined as the spherical image or characteristic for some family $\mathfrak{A}$. Hence, the First Main Theorem does not make sense for T_{jf} if $0 \leqq j < s$. If

$j = 0$, then

$$X(r) = A_{0,f}(r) = \int_{G_r} \chi_m$$

is nothing else but the volume of G_r in the semi-Kaehler metric χ_1 on M, and is independent of f.

Because A is compact, a constant $c > 0$ exists such that (Lemma 7.3)

$$0 \leqq \hat{\lambda} \leqq c\omega_{k-1}$$

on M. Then $\pi^*(c\omega_{k-1} - \hat{\lambda}) \geqq 0$ and $\tau_*\pi^*(c\omega_{k-1} - \hat{\lambda}) \geqq 0$. Hence,

$$0 \leqq \hat{\Lambda} \leqq c\Omega_{s-1}$$

on N. Therefore, $f^*(\hat{\Lambda}) \leqq cf^*(\Omega_{s-1})$ on M. This implies

$$\begin{aligned} \Delta_f(r) &= \int_{G_r-g} f^*(\hat{\Lambda}) \wedge d^{\perp}dh \wedge \chi_q \\ &= \int_{G_r-g} f^*(\hat{\Lambda}) \wedge \chi_{q+1} \\ &\leqq c\int_{G_r} f^*(\Omega_{s-1}) \wedge \chi_{q+1} = cA_{s-1,f}(r). \end{aligned}$$

Therefore, Theorem 9.7 implies

<u>Theorem 9.8. Assume (A1)-(A5) and (A7)-(A9) on the Levi manifold (M,h). If</u>

$$\frac{T_{s-1,f}(r)}{T_{s,f}(r)} = \frac{A_{s-1,f}(r)}{T_{s,f}(r)} \longrightarrow 0 \text{ for } r \to \infty$$

<u>then $b_f = 1$, i.e., $f(M) \cap S_a \neq \emptyset$ for almost all $a \in A$.</u>

A <u>Levi-manifold</u> (M,h) is said to have <u>finite volume</u> if and only if $X(M) = \int_M \chi_m < \infty$. Because $A_{0f}(r) = X(r) \leqq X(M)$ and $T_f(r) \to \infty$ if $r \to \infty$, Theorem 9.8 implies

<u>Theorem 9.9.</u>[23] <u>Assume (A1)-(A5) and (A7)-(A9) on the Levi manifold (M,h). If s = 1 and if (M,h) has finite volume, then $f(M) \cap S_a \neq \emptyset$ for almost every $a \in A$.</u>

Now, two examples shall be given to Case IV. <u>The first is due to Hirschfelder [7]:</u>

Let M be a closed complex submanifold of $\mathbb{C}^j$ with $0 \notin M$. Define $h: M \to R$ by

$$h(z) = \frac{1}{4} \log |z|^2 + c \text{ for } z \in M$$

where the constant $c > 0$ can be chosen such that $h(z) \geqq 0$ on M. Obviously, $h: M \to \mathbb{R}$ is proper and has class C^∞. If $w: U \to M$ is a holomorphic map, of an open subset U of $\mathbb{C}$ into M with $dw \neq 0$, then

$$w^*(d^\perp dh) = |z \wedge w'|^2 \, \frac{1}{2} \, du \wedge d\bar{u} \, .$$

Hence $d^\perp dh > 0$ at every point $z \in M$ whose tangent-plane does not contain 0, i.e., $d^{\perp}dh > 0$ outside a thin analytic subset of M. Moreover $d^{\perp}dh \geq 0$ on M. Hence (M,h) is a Levi manifold. Hence Theorem 9.8 applies.

In [25], it was proven that M is algebraic if and only if the volume $X(M) = \int_M (d^{\perp}dh)^m$ is finite. Hence, if M is algebraic, then (M,h) is a Levi manifold with finite volune. Theorem 9.9 applies.

For the second example compare [30]: Let V be a complex vector space of dimension n + 1 with $0 < n < \infty$. Adopt the same notations as in §2, page 18. Take a (positive definite) hermitian product on V. Then a hermitian product is induced on $\bigwedge_{p+1} V = V[p+1]$, for $p = 0,\cdots,n-1$. This defines a Kaehler metric on $\mathbb{P}(V[p+1])$ with fundamental form $\omega_{p,1}$ such that

$$\mathbb{P}^*(\omega_{p,1})(z) = \frac{1}{2\pi} d^{\perp}d \log |z|$$

for $0 \neq z \in V[p+1]$. Now, the Grassmann-manifold $G_p(V)$ is an algebraic submanifold of $\mathbb{P}(V[p+1])$ with dimension $d_p = (p+1)(n-p)$. Denote the restriction of $\omega_{p,1}$ to $G_p(V)$ by $\omega_{p,1}$ again. As usually, define

$$\omega_{pj} = \omega_{p,1} \wedge \cdots \wedge \omega_{p,1} \qquad \text{(j times)}.$$

If $d = d_p$, abbreviate $\omega_{p,d} = \omega_{[p]}$ and $\omega_{p,d-j} = \omega_{[p]-j}$. Let W(n,p) be the volume of $G_p(V)$:

$$W(n,p) = \int_{G_p(V)} \omega_{[p]} > 0,$$

because $\omega_{[p]} > 0$ on $\mathbb{P}(V[p+1])$.

Lemma 9.10. $W(n,0) = 1$

Proof. Define $T(z) = \frac{1}{4}\, d^{\perp} d|z|^2 = \frac{i}{2}\, \partial\bar{\partial}|z|^2$. Then

$$\tilde{\omega}_{0,n} = \mathbb{P}^*(\omega_{0,n}) = \left(\frac{i}{2\pi}\, \partial\bar{\partial} \log |z|^2\right)^n$$

$$= \frac{1}{\pi^n} \frac{1}{|z|^{4n}} \left(|z|^2 \upsilon - \partial|z|^2 \wedge \bar{\partial}|z|^2\right)^n$$

$$= \frac{1}{\pi^n} \left(\frac{\upsilon^n}{|z|^{2n}} - n\, \partial|z|^2 \wedge \bar{\partial}|z|^2\, \frac{\upsilon^{n-1}}{|z|^{2n+2}}\right).$$

Define φ by

$$\varphi(z) = d^{\perp} \log \frac{1}{|z|} \wedge \upsilon^n(z)$$

Let S_r be the sphere of radius r in V. According to Appendix I Lemma A I 16

$$\int_{S_r} \varphi = 2\pi^{n+1} r^{2n}.$$

Now

$$d|z|^2 \wedge d^{\perp} \log \frac{1}{|z|} = \frac{i}{2} \left(\partial|z|^2 + \bar{\partial}|z|^2\right) \wedge \left(\bar{\partial}|z|^2 - \partial|z|^2\right)|z|^{-2}$$

$$= i|z|^{-2} \partial|z|^2 \wedge \bar{\partial}|z|^2.$$

Therefore,

$$\tilde{\omega}_{0n} \wedge d|z|^2 \wedge d^{\perp} \log \frac{1}{|z|} = \pi^{-n}|z|^{-2n} d|z|^2 \wedge \varphi(z).$$

Therefore, as in Lemma A I 16,

$$J = \int_V e^{-|z|^2} \tilde{\omega}_{0n} \wedge d|z|^2 \wedge d^{\perp} \log \frac{1}{|z|}$$

$$= \pi^{-n} \int_0^{\infty} e^{-r^2} r^{-2n} \int_{S_r} \varphi \, dr^2$$

$$= 2\pi \int_0^{\infty} e^{-r^2} dr^2 = 2\pi.$$

By integration over the fibers of $\mathbb{P}$:

$$J = i \int_{\mathbb{P}(V)} \left(\int_{\mathbb{C}} e^{-|z|^2} |z|^{-2} \partial |z|^2 \wedge \bar{\partial} |z|^2 \right) \omega_{0n}$$

$$= i \int_{\mathbb{P}(V)} \left(\int_{\mathbb{C}} e^{-|z|^2} \partial z \wedge \partial \bar{z} \right) \omega_{0n}$$

$$= 2\pi \int_{\mathbb{P}(V)} \omega_{0n} = 2\pi W(n,0).$$

Therefore, $W(n,0) = 1$, q.e.d.

$W(n,p)$ is the degree of the Grassmann manifold

$$W(n,p) = d_p! \frac{p!(p-1)! \cdots 1!}{(n-p)!(n-p+1)! \cdots n!}$$

However, this will not be needed here.

If $a \in G_p(V)$, then $E(\alpha)$ is a $(p+1)$-dimensional subspace of V. Its associate projective space is $\ddot{E}(a)$. The hermitian product on V, defines a hermitian product on $E(\alpha)$, whose Kaehler matric on $\ddot{E}(\alpha)$ is given by the restriction of $\omega_{0,1}$ to $\ddot{E}(\alpha)$. Hence

$$\int_{\ddot{E}(a)} \omega_{0p} = 1$$

Now, consider the admissible family $\mathfrak{A}_p(V)$ of §2, Example 3, which is given by the triplet

$$\mathbb{P}(V) \xleftarrow{\tau} F_{0,p} \xrightarrow{\pi} G_p(V)$$

where $E(a) = \tau\pi^{-1}(a)$ for $a \in G_p(V)$. Therefore

$$\pi_*\tau^*(\omega_{0,p}) = 1.$$

For $j = 1,\cdots,s = n - p$ define

$$\Omega_{p,j} = \tau_*\pi^*(\omega_{[p]-s+j}) \geqq 0$$

Obviously, $\Omega_{p,j}$ is invariant under the isometries of $\mathbb{P}(V)$. Hence,[24] constants $c_{pj} \geqq 0$ exist such that

$$\Omega_{p,j} = c_{p,j}\omega_{0j} \qquad \text{on } \mathbb{P}(V).$$

Lemma 9.11. $c_{p,s} = W(n,p)$ with $s = n - p$.

Proof. It is

$$c_{pj} = \int_{\mathbb{P}(V)} c_{pj}\omega_{On} = \int_{\mathbb{P}(V)} \Omega_{ps}\omega_{O,p}$$

$$= \int_{F_{O,p}} \pi^*(\omega_{[p]}) \wedge \tau^*(\omega_{O,p})$$

$$= \int_{G_p(V)} \omega_{[p]} \wedge \pi_*\tau^*(\omega_{O,p}) = W(n,p)$$

q.e.d.

Let (M,h) be a Levi manifold of dimension m with $0 \leqq q = m - s \leqq m$ where $s = n - p$. Let $\lambda_{[p]}$ be a proper proximity from to $\frac{1}{W(n,p)}\omega_{[p]}$ on $G_p(V)$ for the point family on $G_p(V)$. Let $f: M \to N$ be a holomorphic map which is almost adopted to $\mathfrak{A}_p(V)$ and $\mathfrak{A}_{p+1}(V)$. Then (A1)-(A5), (A7)-(A9) can be satisfied for $\mathfrak{A}_p(V)$ and $\mathfrak{A}_{p+1}(V)$. The characteristic and the spherical image for $\mathfrak{A}_p(V)$ are

$$T_f(r,p) = \frac{1}{W(n,p)}\int_{G_r} \psi_r f^*(\Omega_{p,s}) \wedge \chi_q$$

$$A_p(r,p) = \frac{1}{W(n,p)}\int_{G_r} f^*(\Omega_{p,s}) \wedge \chi_q$$

because the volume element on $G_p(V)$ has to be normalized. Hence

$$T_f(r,p) = \int_{G_r} \psi_r f^*(\omega_{O,s}) \wedge \chi_q$$

$$A_f(r,p) = \int_{G_r} f^*(\omega_{O,s}) \wedge \chi_q$$

If $d = d_p$ define $a_j = W(n,p)^{-\frac{d-s+j}{a}}$ for $j = 0,\cdots,s$

$$A_{jf}(r,p) = a_j \int_{G_r'} f^*(\Omega_{pj}) \wedge \chi_{m-j} .$$

$$T_{jf}(r,p) = a_j \int_{G_r} f^*(\Omega_{pj}) \wedge \chi_{m-j} .$$

for $j = 0,\cdots,s$. Then

$$A_{jf}(r) = C_{pj} a_j A_f(r,n-j)$$

Therefore, <u>Theorem 9.8 reads:</u> <u>If</u>

$$\frac{T_f'(r,p+1)}{T_f(r,p)} = \frac{A_f(r,p+1)}{T_f(r,p)} \longrightarrow 0 \qquad \text{for } r \to \infty$$

<u>then f(M) intersects almost all p-dimensional projective linear subspaces of $\mathbb{P}(V)$. Therefore, if the derivative of the characteristic of the (p+1)-dimensional projective linear subspaces of $\mathbb{P}(V)$ is "small" relative to the characteristic of the p-dimensional projective linear subspaces of $\mathbb{P}(V)$, then the image f(M) intersects "many" p-dimensional linear projective subspaces.</u> This gives a quite geometric interpretation of the Chern's equidistribution condition [25] for a map of a Levi manifold into the projective space.

Appendix I

The existence and continuity of certain integrals.

Certain highly specialized and complicated Lemmata shall be proven in this appendix. They are the basis for the proof that a singular potential is a proximity form. The original version of these lemmata were given in [28]. Hirschfelder [6] made these Lemmata dependent on a parameter. The Lemmata A I 2 to A I 8 and their proofs are taken verbatim from Hirschfelder [6], and are reproduced here only for the convenience of the reader because [6] is not easily accessible and some of these are not reproduced in [7]. A I 11 to A I 18 are generalizati rs of result of [28].

The following situation is considered:

Situation AI 1.[26)]

(S1): Let Y be an open, relative compact neighborhood of the point c of the complex manifold Y_0.

(S2): Let U and V be open neighborhoods of 0 in $\mathbb{C}^n$ with $\overline{V}$ compact and contained in U.

(S3): A holomorphic map $h: U \times Y_0 \to \mathbb{C}^n$ is given. For each $y \in Y_0$ define $h_y: U \to \mathbb{C}^n$ by $h_y(x) = h(x,y)$ if $x \in U$. Define $\tilde{h}: U \times Y_0 \to \mathbb{C}^n \times Y_0$ by $h(x,y) = (h(x,y),y)$.

(S4): Assume that $h_y: U \to h_y(U)$ is biholomorphic for each $y \in Y_0$.

(S5): Suppose that $h_e(x) = x$ for all $x \in U$.

(S6): Suppose that $h_y(V) \subseteq U$ for all $y \in \overline{Y}$.

(S7): Let $M_0 \neq \emptyset$ be open in $\mathbb{C}^m$ with $m - n = q \geqq 0$. Let $f: M_0 \to U$ be an open, holomorphic map. Define $M = f^{-1}(V)$.

(S8): Define $F: M \times Y \to U \times Y$ by $F(z,y) = \tilde{h}(f(z),y)$. Define $|F|: M \times Y \to \mathbb{R}$ by $|F|(z,y) = |h(f(z),y)|$ and write $|F(z,y)| = |F|(z,y)$.

(S9): For every compact subset K of M, for every $y \in Y$ and for every $\rho \in \mathbb{R}$ with $0 < \rho \leqq 1$ define

$$L(\rho) = L_y(\rho,K) = \{z \in K \mid \frac{\rho}{2} \leqq |F(z,y)| \leqq \rho\}.$$

(S10): On $M \times Y$ define

$$E(k,s) = \left(\log \frac{1}{|F|}\right)^k \frac{1}{|F|^s},$$

if k and s are nonnegative integers.

For every set X denote by θ_X the projection $\theta_X: X \times Y_0 \to X$. For $y \in Y_0$ define $j_y: M \to M \times Y$ by $j_y(z) = (z,y)$. If $(z_1,\ldots,z_n) = z$ are the coordinates on $\mathbb{C}^n$ if $\varphi \in T(q,n)$ and $\psi \in T(q,n)$, write

$$dz_\varphi = dz_{\varphi(1)} \wedge \dots \wedge dz_{\varphi(q)}$$

$$dz_{\varphi\psi} = (\tfrac{i}{2})^q dz_{\varphi(1)} \wedge d\bar{z}_{\psi(1)} \wedge \dots \wedge dz_{\varphi(q)} \wedge d\bar{z}_{\psi(q)}$$

$$= \frac{(i)^{q^2}}{2^q} dz_\varphi \wedge \overline{dz_\psi} = d\bar{z}_{\psi\varphi} = \overline{dz}_{\psi\varphi} .$$

Then $dz_{\varphi\varphi} \geqq 0$ and

$$0 < \upsilon_q = \sum_{\varphi\in T(q,n)} dz_{\varphi\varphi} = \frac{1}{q!} \underbrace{\upsilon_1 \wedge \dots \wedge \upsilon_q}_{(q\text{-times})}$$

<u>Lemma AI.2.</u>[27] Assume Situation AI 1. Suppose that s is an integer with $0 \leqq s < n$. Let K be a compact subset of M. Let χ be a locally bounded form of bidegree (m-s,m-s) on M x Y. Suppose that $j_y(\chi)$ is measurable for each $y \in Y$. Take $\varphi \in T(s,n)$. If $\rho \in \mathbb{R}$ with $0 < \rho \leqq 1$ and if $y \in Y$ define

$$I_y(\rho) = \int_{L_y(\rho,K)} |j_y^*(E(k,2s)F^*\theta_u^*(dz_{\varphi\varphi}) \wedge \chi)|$$

<u>Then $I_y(\rho) \to 0$ for $\rho \to 0$ uniformly on every compact subset of Y.</u>

<u>Proof:</u> <u>Step A:</u> Without loss of generality, $\varphi(\nu) = \nu$ for $\nu = 1,\dots,s$ can be assumed. Write $\theta_u \circ F = (f_1,\dots,f_n)$ where $f_i: M \times Y \to \mathbb{C}$ is holomorphic. Then

$$F^*\theta^*_u(dz_{\varphi\varphi}) \wedge \chi = (\tfrac{i}{2})^s df_1 \wedge d\bar{f}_1 \wedge \dots \wedge df_s \wedge d\bar{f}_s$$

An open neighborhood H of K exists such that $\overline{H}$ is compact, $\overline{H} \subset M$, and such that H is the finite union of balls. Now,

$$\chi = \sum_{\alpha,\beta\in T(m-s,m)} \chi_{\alpha\beta}\, \theta_u^* dz_{\alpha\beta} + \xi$$

with $j_y^*\xi = 0$ for each $y \in Y$. Let Y_1 be a compact subset of Y. A constant $B > 0$ exists such that $|\chi_{\alpha\beta}| \leqq B$ on $K \times Y_1$. For $\alpha \in T(m-s,m)$, define

$$F_\alpha = \frac{\partial(f_1,\ldots\ldots\ldots\ldots,f_s)}{\partial(z_{\alpha^*(m-s+1)},\ldots,z_{\alpha^*(m)})}$$

where $\alpha^*: \Delta_m \to \Delta_m$ is the bijective map such that $\alpha^*|\Delta_{m-s} = \alpha$ and $\alpha^*(\mu) < \alpha^*(\mu+1)$ if $m - s < \mu < \mu + 1 \leqq m$. If $s = 0$, set $F_\alpha = 1$. If $y \in Y_1$, then

$$|j_y^*(F^*\theta_u^*(dz_{\varphi\varphi}) \wedge \chi)|$$

$$= |\sum_{\alpha,\beta\in T(m-s,m)} (\text{sign}\alpha^*\text{sign}\beta^*)(F_\alpha \circ j_y)(F_\beta \circ j_y)(\chi_{\alpha\beta} \circ j_y)|\upsilon_m$$

$$\leqq |\frac{B}{2} \sum_{\alpha,\beta\in T(m-s,m)} (|F_\alpha \circ j_y|^2 + |F_\beta \circ j_y|^2)\upsilon_m$$

$$= \binom{m}{s} B\left(\frac{i}{2}\right)^s j_y^*(df_1 \wedge d\overline{f}_1 \wedge \ldots \wedge df_s \wedge d\overline{f}_s) \wedge \upsilon_{m-s}$$

$$= \binom{m}{s} B j_y^*(F^*\theta_u^*(dz_{\varphi\varphi})) \wedge \upsilon_{m-s}.$$

For $\rho \in \mathbb{R}$ with $0 < \rho < 1$ define

$$T_y(\rho) = \{z \in H \mid \tfrac{\rho}{2} \leqq |F(z,y)| \leqq \rho\} \times \{y\}$$

$$J_y(\rho) = \int_{T_y(\rho)} E(k,rs) F^* \theta_u^*(dz_{\varphi\varphi}) \wedge \theta_M^*(\upsilon_{m-s}).$$

Then $j_y(L_y(\rho)) \subseteq T_y(\rho)$. If $y \in Y_1$, then

$$I_y(\rho) \leqq \binom{m}{s} B\, J_y(\rho).$$

Therefore, it suffices to prove $J_y(\rho) \to 0$ for $\rho \to \infty$ uniformly on Y_1.

Step B: Define $\pi: \mathbb{C}^n \to \mathbb{C}^s$ as the projection onto the first s coordinates. If $s = 0$, then $\mathbb{C}^s = \{0\}$. The maps

$$g = (\pi \times \mathrm{Id}) \circ F : M \times Y \to \mathbb{C}^s \times Y$$

has pure fiberdimension $q + n - s = m - s$. Also $dz_{\varphi\varphi} = \pi^* \upsilon_s$ and $F^* \theta_u^* dz_{\varphi\varphi} = g^* \tilde{\theta}^* \upsilon_s$ where $\tilde{\theta} = \theta_{\mathbb{C}^s}$. By [27], Proposition 2.9

$$J_y(\rho) = \int_{w \in \mathbb{C}^s} \left(\int_{z \in g^{-1}(w,y) \cap T_y(\rho)} \nu_{\pi \circ g \circ j_y}(z) E(k,2s)(z,y) \theta_M^* \upsilon_{m-s} \right) \upsilon_s$$

By Draper [4] Proposition 7.2 or [26] Proposition 5.7 $\nu_{\pi \circ g \circ j_y}(z) = \nu_g(z,y)$. If $z \in g^{-1}(w,y) \cap T_y(\rho)$, then

$$|w| = |g(z,y)| \leqq |F(z,y)| \leqq \rho < 1$$

Hence,

$$0 \leq J_y(\rho) = \int_{|w|\leq\rho} \left(\int_{g^{-1}(w,y)\cap T_y(\rho)} \nu_g E(k,2s)\,\theta_M^* \upsilon_{m-s} \right) \upsilon_s$$

$$\leq \frac{4^s}{\rho^{2s}} \int_{|w|\leq\rho} \left(\int_{g^{-1}(w,y)\cap T_y(\rho)} \nu_g E(k,0)\,\theta_M^* \upsilon_{m-s} \right) \upsilon_s$$

$$\leq 4^s \int_{|w|\leq 1} \left(\int_{g^{-1}(\rho w,y)\cap T_y(\rho)} \nu_g E(k,\rho)\,\theta_M^* \upsilon_{m-1} \right) \upsilon_s$$

for $0 < \rho < 1$. A number $Q < 1$ exists such that $|F(z,y)| \leq Q$ if $(z,y) \in \overline{H} \times Y_1$. Then

$$E(k,0) = \left(\log \frac{1}{|F|}\right)^k \leq \left(\log \frac{Q}{|F|}\right)^k .$$

Define

$$G_\rho(w,y) = \int_{g^{-1}(\rho w,y)\cap T_y(\rho)} \nu_g \left(\log \frac{Q}{|F|}\right)^k \theta_M^* \upsilon_{m-s}$$

for $0 < \rho < 1$, $|w| \leq 1$ and $y \in Y_1$. Then

$$J_y(\rho) \leq 4^s \int_{|w|\leq 1} G_\rho(w,y)\upsilon_s .$$

Define $D = \{w \in \mathbb{C}^s \,|\, |w| \leq 1\}$; then it suffices to prove that $G_\rho(w,y) \to 0$ for $\rho \to 0$ uniformly on $D \times Y_1$.

<u>Step C</u>: The following statement shall be proved: For $(z,y) \in M \times Y$ define $\delta(z,y)$ to be the distance from z to the compact set $\overline{H} \cap \theta_M(F^{-1}(0,y)) = \overline{H} \cap f^{-1}(h_y^{-1}(0))$. For $\varepsilon > 0$

$$A(\varepsilon) = \{(z,y) \in M \times Y_1 | \delta(z,y) < \varepsilon\}$$

$$A_y(\varepsilon) = \{z \in M | (z,y) \in A(\varepsilon)\} \qquad \text{if } y \in Y_1.$$

Then $\varepsilon_0 > 0$ exists such that $0 < \varepsilon' < \varepsilon \leqq \varepsilon_0$ implies $\overline{A(\varepsilon')} \subset A(\varepsilon) \subset \overline{A(\varepsilon)} \subset M \times Y_1$ and $\overline{A(\varepsilon)}$ is compact.

Proof: Let H_0 be an open neighborhood of $\overline{H}$ with compact $\overline{H}_0 \subset M$. Let $\delta'(z)$ be the distance from z to $\overline{H}$. Then $\varepsilon_0 > 0$ exists such that $\{z | \delta'(z) \leqq \varepsilon_0\} \subset H_0$. Now, $\delta'(z) \leqq \delta(z,y)$ for each $y \in Y_1$ implies $\overline{A(\varepsilon_0)} \subseteq H_0 \times Y_1$. Hence $\overline{A(\varepsilon_0)}$ is compact.

Suppose $0 < \varepsilon' < \varepsilon \leqq \varepsilon_0$. Take $(z,y) \in A(\varepsilon')$. Let $\{(z_\nu, y_\nu)\}_{\nu \in \mathbb{N}}$ be a sequence of points in $A(\varepsilon')$ converging to (z,y) for $\nu \to \infty$. For each $\nu \in \mathbb{N}$, a point $\mu_\nu \in \overline{H} \cap \theta_M F^{-1}(0,y_\nu)$ with $|z_\nu - \mu_\nu| < \varepsilon'$ exists. Choose a convergent subsequence $\mu_{\nu_\lambda} \to \mu \in \overline{H}$ for $\lambda \to \infty$. Then $|z-\mu| \leqq \varepsilon'$. Since $h(f(\mu_{\nu_\lambda}), y_{\nu_\lambda}) = 0$, also $h(f(\mu),y)=0$, that is, $\mu \in \overline{H} \cap \theta_M F^{-1}(0,y)$. Hence $\delta(z,y) \leqq \varepsilon' < \varepsilon$ and $(z,y) \in A(\varepsilon)$, which proves Statement C.

Step D: The following statement shall be proved: For $0 < \varepsilon \leqq \varepsilon_0$, a number $\rho_0(\varepsilon)$ with $0 < \rho_0(\varepsilon) < 1$ exists such that $T_y(\rho) \subset A(\varepsilon)$ for all $y \in Y_1$ and $0 < \rho < \rho_0(\varepsilon)$.

Proof. Suppose the statement were wrong. A sequence $\{\rho_\nu\}_{\nu \in \mathbb{N}}$ with $0 < \rho_\nu < 1$ and a sequence $\{(z_\nu y_\nu)\}_{\nu \in \mathbb{N}}$ exist such that $\rho_\nu \to 0$ for

$\nu \to \infty$ and $(z_\nu, y_\nu) \in T_{y_\nu}(\rho) - A(\varepsilon)$ with $y_\nu \in Y_1$. Because $(z_\nu, y_\nu) \in \overline{H} \times Y_1$, it can be assumed that $(z_\nu, y_\nu) \to (z,y)$ for $\nu \to \infty$. Now $\frac{\rho_\nu}{2} \leq |F(z_\nu, y_\nu)| \leq \rho_\nu$ implies $F(z,y) = (0,y)$ or $z \in \theta_M F^{-1}(0,y)$. Let W be an open neighborhood of y in Y. Set $D_\varepsilon = \{z' \in M \mid |z'-z| < \frac{\varepsilon}{2}\}$. Because $F: M \times Y \to U \times Y$ is open, $F(D_\varepsilon \times W)$ is open in $U \times Y$. Hence ν_0 exists such that $(0, y_\nu) \in F(D_\varepsilon \times W)$ if $\nu \geq \nu_0$. Therefore $z'_\nu \in D_\varepsilon$ exists such that $F(z'_\nu, y_\nu) = (0, y_\nu)$. Hence $z'_\nu \in \theta_M F^{-1}(0, y_\nu)$ with $|z'_\nu - z| < \frac{\varepsilon}{2}$ if $\nu \geq \nu_1$. Hence $|z_\nu - z'_\nu| < \varepsilon$ if $\nu \geq \nu_1$. This means $(z_\nu, y_\nu) \in A(\varepsilon)$ if $\nu \geq \nu_1$, which is a contradiction. Hence, Statement D is true.

<u>Step E</u>: $G_\rho(w,y) \to 0$ for $\rho \to 0$ uniformly on $D \times Y_1$ shall be proved now:

Take a C^∞-function λ on $\mathbb{R}$ with $0 \leq \lambda \leq 1$ and with $\lambda(x) = 1$ if $x < \frac{1}{2}$ and $\lambda(x) = 0$ if $x > \frac{3}{4}$. For $0 < \varepsilon \leq \varepsilon_0$ define $\lambda_\varepsilon: M \times Y_1 \to \mathbb{R}$ by $\lambda_\varepsilon(z,y) = \lambda(\frac{1}{\varepsilon}\delta(z,y))$. Then λ_ε is continuous and has compact support in $A(\varepsilon)$. Moreover $\lambda_\varepsilon(z,y) \to 0$ for $\varepsilon \to 0$, on $M \times Y_1 - F^{-1}(0,y)$. For $0 < \varepsilon \leq \varepsilon_0$ and $y \in Y_1$ define

$$0 \leq \hat{G}^\varepsilon(y) = \int_{g^{-1}(0,y) \cap (\overline{H} \times Y_1) \cap A(\varepsilon)} \nu_g (\log \frac{Q}{|F|})^k \cdot \theta_M^* \upsilon_{m-s}$$

Now, $\bigcap_{\varepsilon > 0} A(\varepsilon) \cap (M \times \{y\}) = F^{-1}(0,y) \cap \overline{H}$ is a set of measure zero on $g^{-1}(0,y)$; therefore, $\hat{G}^\varepsilon(y) \to 0$ for $\varepsilon \to 0$.

Take $\eta > 0$. Then $\varepsilon_1(y,\eta)$ with $0 < \varepsilon_1(y,\eta) \leqq \varepsilon_0$ exists such that $\hat{G}^{\varepsilon}(y) < \frac{\eta}{2}$ if $0 < \varepsilon < \varepsilon_1(y,\eta)$. Define

$$G^{\varepsilon}(w,y) = \int_{g^{-1}(w,y)\cap(\overline{H}\times Y_1)} \nu_g(\log \frac{Q}{|F|})^k \cdot \lambda_\varepsilon \cdot \theta_M^* \upsilon_{m-s}$$

for $|w| \leqq 1$ and $y \in Y$. By [27] Theorem 4.9, this function is continuous at $(0,y_0)$ for every $y_0 \in Y$. Hence, $\varepsilon_2(y_0,\varepsilon,\eta)$ and a neighborhood $W(y_0,\varepsilon,\eta)$ of $y_0 \in Y_1$ exists such that

$$|G^{\varepsilon}(w,y) - G^{\varepsilon}(0,y_0)| < \frac{\eta}{2}$$

if $|w| < \varepsilon_2(y_0,\varepsilon,\eta)$ and $y \in W(y_0,\varepsilon,\eta)$. If $0 < \rho < \rho_0(\frac{\varepsilon}{2})$, then $G_\rho(w,y) \leqq G^{\varepsilon}(\rho w,y)$. Take $\varepsilon = \varepsilon_1(y_0,\eta)$ and $\rho(y_0,\eta) = \operatorname{Min}(\rho_0(\frac{\varepsilon}{2}), \varepsilon_2(y_0,\tau,\eta))$. Take ρ with $0 < \rho < \rho(y_0,\eta)$. Then

$$G_\rho(w,y) \leqq G^{\varepsilon}(\rho w,y) \leqq G^{\varepsilon}(0,y_0) + \frac{\eta}{2}$$

$$\leqq \hat{G}^{\varepsilon}(y_0) + \frac{\eta}{2} < \eta$$

if $|w| \leqq 1$ and $y \in W(y_0,\varepsilon_1(y_0,\eta),\eta) = W_{y_0}(\eta)$. Finitely many points $y_{01},\ldots,y_{0p}$ exist in Y_1 such that

$$Y_1 \subseteq \bigcup_{i=1}^{p} W_{y_{0p}}(\eta).$$

Define $\rho_1(\eta) = \mathrm{Min}(\rho(y_{01}, \eta),\ldots,\rho(y_{0,p})\eta))$. Then $0 \leqq G_\rho(w,y) < \eta$ if $\rho < \rho_1(\eta)$ for all $(w,y) \in D \times Y_1$; q.e.d.

Lemma A I 3.[28) Assume Situation A I 1. Let K be a compact subset of M. Let χ be a locally bounded form of bidegree (q,q) on $M \times Y$. Suppose that $j_y(\chi)$ is measurable for each $y \in Y$. For $\rho \in \mathbb{R}$ with $0 < \rho < 1$ and $y \in Y$ define

$$I_y(\rho) = \int_{L_y(\rho)} |j_y^*(E(0,2n)F^*\theta^*\upsilon_n \wedge \chi)| .$$

Then $I_y(\rho)$ is uniformly bounded on compact subsets of Y.

Proof. Case 1: $q > 0$. Let $\varphi \in T(q,q)$ be the identity. Construct $H, B, T_y(\rho), J_y(\rho) dz_{\varphi\varphi} = \upsilon_n, \pi = \mathrm{Id}\colon \mathbb{C}^p \to \mathbb{C}^p$ and $g = F$ as in the proof of the last Lemma with $s = n$ and $k = 0$. Then

$$I_y(\rho) \leqq \binom{m}{n} B\, J_y(\rho)$$

where

$$\begin{aligned} J_y(\rho) &= \int_{T_y(\rho)} \frac{1}{|F|^{2n}} F^*\theta_U^*\upsilon_n \wedge \theta_M^*\upsilon_q \\ &= \int_{\frac{\rho}{2}\leqq|w|\leqq\rho} \frac{1}{|w|^{2p}} \Big(\int_{g^{-1}(w,y)\cap T_y(\rho)} \nu_g \theta_M^*\upsilon_q\Big)\upsilon_n \\ &\leqq 4^n \int_{\frac{1}{2}\leqq|w|\leqq 1} \Big(\int_{g^{-1}(\rho w,y)\cap(\overline{H}\times Y)} \nu_g\theta_M^*\upsilon_q\Big)\upsilon_n . \end{aligned}$$

Let Y_1 be compact in Y. By [27] Theorem 4.9 the inside integral is a continuous function of ρw and y on $\{(\rho,w) \mid |\rho w| \leqq 1\} \times Y_1$, hence bounded by a constant D on this set. Therefore

$$0 \leqq J_y(\rho) \leqq \frac{(4\pi)^n}{n!} D$$

for $0 < \rho \leqq 1$ and $y \in Y_1$.

<u>Case 2.</u> $q = 0$. Then $n = m$ and F is an open, light map. Let Y_1 be compact in Y. Then $K \times Y_1$ is compact. By [26], Lemma 2.5 a constant $D > 0$ exists such that

$$\sum_{z \in K \times Y_1} \nu_F(z;w,y) \leqq D$$

if $w \in \mathbb{C}^p$ and $y \in Y_1$, where $\nu_f(z;w,y) = \nu_p(z)$ if $F(z) = (w,y)$ and $\nu_f(z;w,y) = 0$ if $F(z) \neq (w,y)$.

A constant $B > 0$ exists such that $|\chi(z,y)| \leqq B$ for $z \in K$ and and $y \in Y_1$. Therefore,

$$\sum_{z \in K \times Y_1} \nu_F(z;w,y)\,|\chi(z)| \leqq DB .$$

[27] Lemma 2.8 implies

$$I_y(\rho) = \int_{J_y(L_y(\rho))} |\chi| \, \frac{1}{|F|^{2m}} |F^* \theta_M^* \upsilon_m|$$

$$= \int_{\frac{\rho}{2} \leq |w| \leq \rho} \Big(\sum_{z \in K \times Y_1} \nu_F(z;w,y) \,|\, (z)|\Big) \frac{1}{|w|^{2m}} \upsilon_m$$

$$\leq 4^m DB \frac{\pi^m}{m!}$$

if $y \in Y_1$ and $0 < \rho \leq 1$. q.e.d.

Lemma A I 4.[29] Assume Situation A I 1. Let s and t be integers with $0 \leq s < n$ and $0 \leq t \leq n$. Set $\sigma = m - s$ and $\tau = m - t$. Take $\varphi \in T(s,n)$ and $\psi \in T(t,n)$. Let K be a compact subset of M. Let χ be a locally bounded form of bidegree (σ,τ) on M x Y. Suppose that $j_y(\chi)$ is measurable for each $y \in Y$. For $\rho \in \mathbb{R}$ with $0 < \rho \leq 1$ define

$$I_y(\rho) = \int_{L_y(\rho)} |j_y^*(E(k,s+t)F^*\theta_U^*(dz_\varphi \wedge d\overline{z}_\psi) \wedge \chi)|.$$

If $\alpha \in T(\sigma,m)$ and $\beta \in T(\tau,m)$, define

$$I_{y,\alpha}(\rho) = \int_{L_y(\rho)} j_y^*(E(2k,2s)F^*\theta_U^* dz_{\varphi\varphi} \wedge \theta_M^* dz_{\alpha\alpha})$$

$$J_{y,\beta}(\rho) = \int_{L_y(\rho)} j_y^*(E(0,2t)F^*\theta_U^* dz_{\psi\psi} \wedge \theta_M^* dz_{\beta\beta}).$$

Let Y_1 be a compact subset of Y. Then a constant $B > 0$ exists such that $\rho \in \mathbb{R}$ with $0 < \rho \leq 1$ and $y \in Y_1$ imply

$$0 \leq I_y(\rho) \leq B \sum_{\alpha \in T(\sigma,m)} \sum_{\beta \in T(\tau,m)} I_{y,\alpha}(\rho)^{\frac{1}{2}} J_{y,\beta}(\rho)^{\frac{1}{2}}$$

<u>Proof.</u> Write $\theta_M \circ F = (f_1, \ldots, f_n)$ where f_i is holomorphic on $M \times Y$. Then

$$F^* \theta_U^* dz_\varphi = df_\varphi = df_{\varphi(1)} \wedge \cdots \wedge df_{\varphi(s)}$$

$$F^* \theta_U^* d\overline{z}_\psi = d\overline{f}_\psi = d\overline{f}_{\psi(1)} \wedge \cdots \wedge d\overline{f}_{\psi(t)} .$$

Write

$$\chi = \sum_{\alpha \in T(\sigma,m)} \sum_{\beta \in T(\tau,m)} \chi_{\alpha\beta} dz_\alpha \wedge d\overline{z}_\beta + \zeta$$

where $j_y^*(\zeta) = 0$ for each $y \in Y$. A constant $B > 0$ exists such that $|\chi_{\alpha\beta}| \leqq B$ on $K \times Y_1$. If $\gamma \in T(a,b)$ with $a \leqq b$, let $\gamma^* : \Delta_b \to \Delta_b$ be that bijective map such that $\gamma^* | \Delta_a = \gamma$ and $\gamma^*(x) < \gamma^*(x+1)$ if $a < x < b$. Define

$$F_\alpha = \frac{\partial(f_{\varphi(1)}, \ldots\ldots, f_{\varphi(s)})}{\partial(z_{\alpha^*(\sigma+1)}, \ldots, z_{\alpha^*(m)})}$$

$$G_\beta = \frac{\partial(f_{\psi(1)}, \ldots\ldots, f_{\psi(t)})}{\partial(z_{\beta^*(\tau+1)}, \ldots, z_{\beta^*(m)})}$$

for $\alpha \in T(\sigma,m)$ and $\beta \in T(\tau,m)$. Then

$$|j_y^*(F^* \theta_U^*(dz_\varphi \wedge d\overline{z}_\psi) \wedge \chi)|$$

$$= |\sum_{\alpha \in T(\sigma,m)} \sum_{\beta \in T(\tau,m)} \text{sign } \alpha \text{ sign } \beta \, F_\alpha \overline{G}_\beta \chi_{\alpha\beta}| v_m$$

$$\leqq B \sum_{\alpha \in T(\sigma,m)} \sum_{\beta \in T(\tau,m)} |F_\alpha| |G_\beta| v_m .$$

Observe

$$I_{y,\alpha}(\rho) = \int_{L_y(\rho)} E(2k,2s)\,|F_\alpha|^2 \upsilon_m$$

$$I_{y,\beta}(\rho) = \int_{L_y(\rho)} E(0,2s)\,|G_\beta|^2 \upsilon_m$$

Hence,

$$I_y(\rho) \leq B \sum_{\alpha\in T(\sigma,m)} \sum_{\beta\in T(\tau,m)} \int_{L_y(\rho)} E(k,s+t)\,|F_\alpha|\,|G_\beta|\,\upsilon_m$$

$$\leq B \sum_{\alpha\in T(\sigma,m)} \sum_{\beta\in T(\tau,m)} I_{y,\alpha}(\rho)^{\frac{1}{2}}\, J_{y,\beta}(\rho)^{\frac{1}{2}} \qquad \text{q.e.d.}$$

Lemma A I 5.[30] Assume Situation A I.1. Let s and t be integers with $0 \leq s \leq n$ and $0 \leq t \leq n$ where $s + t < 2n$. Set $\sigma = m - s$ $\tau = m - t$. Take $\varphi \in T(s,n)$ and $\psi \in T(t,n)$. Let K be a compact subset of M. Let χ be a locally bounded form of bidegree (τ,σ) on $M \times Y$. Suppose that $J_y(\chi)$ is measurable for each $y \in Y$. For $\rho \in \mathbb{R}$ with $0 < \rho < 1$ and $y \in Y$ define $I_y(\rho)$ as in Lemma A I 4. Then $I_y(\rho) \to 0$ for $\rho \to 0$ uniformly on every compact subset of Y.

Proof. Without loss of generality, $s < n$ can be assumed. Let Y_1 be compact in Y. Then the estimate of Lemma A I 4 holds. Lemma A I 2 implies $I_{y,\alpha}(\rho) \to 0$ for $\rho \to 0$ uniformly on Y_1. If $t < n$, the same Lemma implies $J_{y,\beta}(\rho) \to 0$ for $\rho \to 0$ uniformly on Y_1. If $t = n$, Lemma A I 3 implies that $J_{y,\beta}(\rho)$ is uniformly bounded on Y_1 for $0 < \rho \leq 1$. In either case, $I_y(\rho) \to 0$ for $\rho \to 0$ uniformly on Y_1, q.e.d

Recall the convention about class "C^k" as explained at the beginning of §5 page 52 . So, on a subset of a product space, a form χ is of class $C^\beta(\mu,0)$ if χ has locally bounded coefficients, if each coefficient is measurable in the first variable, for each fixed value of the second, and if each coefficient is continuous in the second variable, for each fixed value of the first.

Lemma A I 6.[31] Assume Situation A I.1. Let K be a compact subset of M. Set $K' = K \times Y - F^{-1}(0,Y)$. Let $\gamma\colon K' \to \mathbb{C}$ be a function of class $C^\beta(\mu,0)$ on K'. For $y \in Y$ define

$$L_y = \{z \in K \mid a \leqq |F(z,y)| \leqq b\}$$

$$\Gamma(y) = \int_{L_y} \gamma(z,y)\upsilon_m(z).$$

Then Γ is continuous on Y.

Proof. Define $L'_y = \{z \in K \mid a < |F(z,y)| < b\}$. Then $\Delta_y = L'_y - L_y$ is compact and has measure zero. Take $y_0 \in Y$. Define

$$L'' = \{z \in K \mid \tfrac{a}{2} \leqq |F(z,y_0)|\}$$

Then $L_{y_0} \subseteq L''$ and L'' is compact. Since F is continuous, a compact neighborhood V of y_0 with $V \subset Y$ exists such that $L_y \subset L''$ for all $z \in V$ and such that $|F(z,y)| \geqq \frac{a}{4} > 0$ if $(z,y) \in L'' \times V$. Hence $L'' \times V$ is a compact subset of K'. Therefore, a constant $C > 0$ exists such that $|\gamma(z,y)| \leqq C$ if $(z,y) \in L'' \times V$.

Define $\lambda_y(z) = 1$ if $z \in L_y$ and $\lambda_y(z) = 0$ if $z \in M - L_y$. Take $z \in L'' - \Delta_{y_0}$. Then $\lambda_y(z) \to \lambda_{y_0}(z)$ for $y \to y_0$ shall be shown. If $z \in L'_{y_0}$, then $a < |F(z,y_0)| < b$. A neighborhood V_0 of y_0 with $V_0 \subseteq V$ exists such that $a < |F(z,y)| < b$ for all $y \in V_0$. Hence $\lambda_y(z) = 1 = \lambda_{y_0}(z)$ for $y \in V_0$. If $z \notin L'_{y_0}$, then $a > |F(z,y_0)|$ or $b < |F(z,y_0)|$, because $z \notin \Delta_{y_0}$. Hence, a neighborhood V_0 of y_0 with $V_0 \subseteq V$ exists such that $a > |F(z,y)|$ or $b < |F(z,y)|$ for all $y \in V_0$. Hence, $\lambda_y(z) = 0 = \lambda_{y_0}(z)$ if $y \in Y_0$. Therefore, $\lambda_y(z) \to \lambda_{y_0}(z)$ for $y \to y_0$ if $z \in L'' - \Delta_{y_0}$.

For $y \in Y$, define $K'_y = \{z \in K | (z,y) \in K'\}$ and $\gamma_y\colon K'_y \to \mathbb{C}$ by $\gamma_y(z) = \gamma(z,y)$. Pick $z \in L''$. Then

$$\gamma_y(z) = \gamma(z,y) \to \gamma(z,y_0) = \gamma_{y_0}(z)$$

for $y \to y_0$ with $y \in V$, because $L'' \times V \subseteq K'$. Hence

$$\lambda_y(z)\gamma_y(z) \to \lambda_{y_0}(z)\gamma_{y_0}(z) \qquad \text{for } y \to y_0$$

if $z \in L'' - \Delta_{y_0}$. Moreover $|\lambda_y(z)\gamma_y(z)| \leqq C$ if $z \in L''$ and $y \in V$. Then

$$\Gamma(y) = \int_{L_y} \gamma(z,y)\upsilon_m = \int_{L''-\Delta_{y_0}} \lambda_y(z)\gamma_y(z)\upsilon_m(z)$$

$$\to \int_{L''-\Delta_{y_0}} \lambda_{y_0}(z)\gamma_{y_0}(z)\upsilon_m = \Gamma(y_0)$$

for $y \to y_0$; q.e.d.

Lemma A I 7.[32] Assume Situation A I.1 holds. Let s and t be integers with $0 \leqq s \leqq n$ and $0 \leqq t \leqq n$ where $s + t < 2n$. Set $\sigma = m - s$ and $\tau = m - t$. Let K be a compact subset of M. Let χ be a locally bounded form of bidegree (σ,τ) on M x Y. Suppose that $j_y^*(\chi)$ is measurable for each $y \in Y$. Take $\varphi \in T(s,n)$ and $\psi \in T(t,n)$. Define

$$\Omega = E(k,s+t)F^*\theta_U^*(dz_\varphi \wedge d\bar{z}_\psi) \wedge \chi .$$

Then $j_y(\Omega)$ and $|J_y(\Omega)|$ are integrable over K. The integrals

$$I(y) = \int_K j_y^*(\Omega) \qquad \text{and} \qquad J(y) = \int_K |j_y^*(\Omega)|$$

are measurable and locally bounded functions on Y.

In addition, if χ is of class $C^\beta(\mu,0)$ on $K' = K \times Y - F^{-1}(0,y)$ then I and J are continuous on Y.

Proof. Again $s < n$ can be assumed. Define

$$K_y' = \{z \in K \mid (z,y) \in K'\}$$

$$K_y^0 = \{z \in K \mid \ |F(z,y)| \geqq \tfrac{1}{2}\} .$$

Then $K_y' = \{z \in K | F(z,y) \neq (0,y)\}$. Hence

$$K_y' = K_y^0 \cup \bigcup_{n=1}^{\infty} L_y(\tfrac{1}{2^n}).$$

Any two sets of this union intersect at most in a set of measure zeros on K. Therefore

$$\int_K |j_y^*\Omega| = \int_{K_y^0} |j_y^*\Omega| + \sum_{n=1}^{\infty} \int_{L_y(\frac{1}{2^n})} |j_y^*\Omega|$$

where each term on the right is a measurable and locally bounded function of y on Y. If χ is of class $C^\beta(\mu,0)$ on K', then, according to Lemma A I 6, each of the terms is a continuous function on Y. Hence, it suffices to prove that the series on the right converges uniformly on each compact subset of Y. Define $I_y(\rho)$ as in Lemma A I 4. Then

$$I_y(\rho) = \int_{L_y(\rho)} |j_y^*\Omega|$$

for $0 < \rho < 1$ and $y \in Y$.

Define $I_{y,\alpha}(\rho)$ and $J_{y,\beta}(\rho)$ as in Lemma A I 4. Let Y_1 be a compact subset of Y. A constant $D_1 > 0$ exists such that $|J_{y,\beta}(\rho)| \leq D_1$ if $0 < \rho \leq 1$ and $y \in Y_1$ and $\beta \in T(\tau,m)$ (Lemma A I 2 if $t < n$, and Lemma A I 3 if $t = n$). Moreover

$$I_{y,\alpha}(\rho) = \int_{L_y(\rho)} j_y^*(\log \tfrac{1}{|F|})^{2k} \frac{1}{|F|^{2s}} F^* \theta_U^* dz_{\varphi\varphi} \wedge \theta_M^* dz_{\alpha\alpha})$$

$$\leqq \frac{1}{(\log \frac{1}{\rho})^4} \int_{L_y(\rho)} j_y^*((\log \frac{1}{|F|})^{2k+4} \frac{1}{|F|^{2s}} F^* \theta_U^* dz_{\varphi\varphi} \wedge \theta_M^* dz_{\alpha\alpha}).$$

By Lemma A I 2 with 2k + 4 in place of k, a constant $D_2 > 0$ exists such that the last integral is bounded by D_2 for $0 < \rho \leqq \frac{1}{2}$ and $y \in Y_1$. Hence

$$0 \leqq I_{y,\alpha}(\rho) \leqq D_2(\log \frac{1}{\rho})^{-4}$$

if $0 < \rho \leqq \frac{1}{2}$ and $y \in Y_1$. Lemma A I 4 implies

$$0 \leqq I_y(\rho) \leqq B \sum_{\alpha\in T(\sigma,m)} \sum_{\beta\in T(\tau,m)} I_{y,\alpha}(\rho)^{\frac{1}{2}} I_{y\beta}(\rho)^{\frac{1}{2}}$$

$$\binom{m}{\sigma}\binom{m}{\tau} B(D_1D_2)^{\frac{1}{2}}(\log \frac{1}{\rho})^{-2} = B_1(\log \frac{1}{\rho})^{-2}$$

where B and B_1 are positive constants. Hence

$$I_y(\frac{1}{2^n}) \leqq B_1 \frac{1}{(\log 2)^2} \frac{1}{n^2}$$

if $y \in Y_1$ and $n \in \mathbb{N}$. Therefore $\sum_{n=1}^{\infty} I_y(\frac{1}{2^n})$ converges uniformly on Y_1, which proves the case for $|j_y^*\Omega|$. The existence of $\int_K |j_y^*\Omega|$ proves the existence of $\int_K j_y^*\Omega$. Then the same proof applies to $j_y^*\Omega$ instead of $|j_y^*\Omega|$, q.e.d.

Lemma A I.8.[33] Assume Situation AI.1. Let s and t be integers $0 \leqq s \leqq n$ and $0 \leqq t \leqq n$ where $s + t < 2n$. Set $\sigma = m - s$ and $\tau = m - t$. Let K be a compact subset of M. Let χ be a locally bounded form of bidegree (σ,τ) on M x Y. Suppose that $j_y^*(\chi)$ is measurable for each $y \in Y$. Let φ be a locally bounded form of bidegree (s,t) on U x Y. For each $y \in Y$, define $i_y: U \to U \times Y$ by $i_y(z) = (z,y)$. Suppose that $i_y^*(\varphi)$ is measurable for each $y \in Y$.

$$\Omega = E(k,s+t)F^*\varphi \wedge \chi.$$

Then $j_y^*\Omega$ and $|j_y^*\Omega|$ are integrable over K and their integrals are measurable and locally bounded functions on Y.

In addition, if χ is of class $C^\beta(\mu,0)$ on $K' = K \times Y - F^{-1}(0,Y)$, on $(U-\{0\}) \times Y$, then the integral

$$I(y) = \int_K j_y^*(\Omega)$$

defines a continuous function I on Y.

Proof. Write

$$\varphi = \sum_{\alpha\in T(s,n)} \sum_{\beta\in T(t,n)} \varphi_{\alpha\beta}\theta_U^*(dz_\alpha \wedge d\bar{z}_\beta) + \zeta$$

where $i_y^*(\zeta) = 0$ for each $y \in Y$. The functions $\varphi_{\alpha\beta}$ are of class $C^\beta(\mu,\beta)$. Then

$$\int_K |j_y^* \Omega| \leq \sum_{\alpha\in T(s,n)} \sum_{\beta\in T(t,n)}$$

$$\int_K |j_y^*(E(k,s+t)(\varphi_{\alpha\beta} \circ F)F^* \theta_U^*(dz_\alpha \wedge d\bar{z}_\beta) \wedge \chi)| .$$

Lemma AI.7 implies (with $(\varphi_{\alpha\beta} \circ F)\,\chi$ instead of χ) that $\int_K |j_y^*\Omega|$ exists and is a locally bounded function of y. Obviously, it is measurable. Hence

$$I(y) = \int_K j_y^*\Omega = \sum_{\alpha\in T(s,n)} \sum_{\beta\in T(t,n)}$$

$$\int_K j_y^*(E(k,s+t)(\varphi_{\alpha\beta} \circ F)F^* \theta_u^*(dz_\alpha \wedge d\bar{z}_\beta) \wedge \chi$$

exists and is a measurable and locally bounded function on Y by Lemma AI.7. Moreover, if the additional assumptions are made, then I is continuous on Y; q.e.d.

Lemma A I.9. Assume Situation AI.1. Let K be a compact subset of M. Let φ be a locally bounded form of degree $r < 2n$ on U x Y. For each $y \in Y$, define i_y: U → U x Y by $i_y(z) = (z,y)$. Suppose that $i_y^*(\varphi)$ is measurable for each $y \in Y$. Let χ be a locally bounded form of degree 2m - r on M x Y. Suppose that $j_y^*(\chi)$ is measurable for each $y \in Y$. Define

$$\Omega = E(k,r)F^*\varphi \wedge \chi .$$

Then $I(y) = \int_K j_y^*(\Omega)$ exists and is a measurable and locally bounded function on Y.

In addition, if χ is of class $C^\beta(\mu;0)$ on $K' = K \times Y - F^{-1}(0,Y)$, and if φ is of class $C^\beta(\mu;0)$ on $(U-\{0\}) \times Y$, then I is continuous on Y.

Proof. Write $\varphi = \sum_{s+t=r} \varphi_{s,t}$ and $\chi = \sum_{\sigma+\tau=2m-r} \chi_{\sigma\tau}$, where $\varphi_{s,t}$ has bidegree (s,t) and $\chi_{\sigma\tau}$ has bidegree (σ,τ). Then

$$F^*\varphi \wedge \chi = \sum_{s+t=r} \sum_{\sigma+\tau=2m-r} F^*(\varphi_{s,t}) \wedge \chi_{\sigma\tau}$$

$$= \sum_{s+t=r} \sum_{s+\sigma=m} \sum_{t+\tau=m} F^*(\varphi_{s,t}) \wedge \chi_{\sigma\tau}$$

$$= \sum_{s+t=r} F^*(\varphi_{s,t}) \wedge \chi_{m-s,m-t}$$

by degree reasons. Now, Lemma AI.8 implies Lemma AI.9, q.e.d.

Theorem AI.10. Let M be a complex manifold of dimension m. Let U and W be non-empty, open subsets of $\mathbb{C}^n$. Suppose that $q = m - n \geqq 0$. Let K be a compact subset of M. Let $f: M \to U$ be an open, holomorphic map. For every $y \in W$, define $j_y: M \to M \times W$ by $j_y(z) = (z,y)$ if $z \in M$ and $i_y: U \to U \times W$ by $i_y(z) = (z,y)$ if $z \in U$.

Let φ be a locally bounded form of degree $r < 2n$ on $U \times W$. Suppose, that $\varphi_y = i_y^*(\varphi)$ is measurable on U for each $y \in W$. Let χ be a locally bounded form of degree $2m - r$ on $M \times W$. Suppose that $\chi_y = j_y^*(\chi)$ is measurable on M for each $y \in W$. Let k be a non-negative integer. Then the integral

$$I(y) = \int_{z \in K} \left(\log \frac{1}{|f(z)-y|}\right)^k \frac{1}{|f(z)-y|^r} f^*(\varphi_y) \wedge \chi_y$$

<u>exists for each $y \in W$. The function I is measurable and locally bounded on W.</u>

<u>If, in addition, φ is of class $C^\beta(\mu,0)$ on $H = \{(x,y) \in U \times W \mid x \neq y\}$ and if χ is of class $C^\beta(\mu;0)$ on $R = \{(z,y) \in M \times W \mid f(z) \neq y\}$, then I is continuous on W.</u>

<u>Proof.</u> Take $y_0 \in W$. It has to be proven that I is locally bounded and measurable, respectively continuous, in a neighborhood of y_0. Without loss of generality $y_0 = 0$ can be assumed.

Define $h\colon \mathbb{C}^n \times W \to \mathbb{C}^n$ by $h(x,y) = x - y$ if $x \in \mathbb{C}^n$ and $y \in W$. Define $h_y\colon \mathbb{C}^n \to \mathbb{C}^n$ by $h_y(x) = h(x,y)$. Then $h_y\colon \mathbb{C}^n \to \mathbb{C}^n$ is biholomorphic and h_0 is the identity. Define $\tilde{h}\colon \mathbb{C}^n \times W \to \mathbb{C}^n \times W$ by $\tilde{h}(x,y) = (h(x,y),y) = (x-y,y)$. Then $\tilde{h}$ is biholomorphic.

Pick any $a \in K$. Then the following statement shall be proven:

<u>Statement.</u> <u>An open neighborhood M_1 of a in M and an open neighborhood W_1 of $y_0 = 0$ in W exist, such that for every C^∞-function $\lambda\colon M \to \mathbb{R}$ with compact support in M_1 the integral</u>

$$I_\lambda(y) = \int_K \left(\log \frac{1}{|f-y|}\right)^k \frac{1}{|f-y|^r} f^*(\varphi_y) \wedge \lambda\chi_y$$

<u>exists for each $y \in W_1$ and defines a locally bounded and measurable, respectively continuous, function I_λ on W_1.</u>

Obviously, if the statement holds, finitely many $M_1^1,\dots,M_1^t$ will cover K. Using a partition of unity on K subjugated to this covering, it follows immediately, that I exists and is a locally bounded and measurable, respectively continuous, function on the open neighborhood $W_1^1 \cap \dots \cap W_1^t$ of $y_0 = 0$. Hence, it remains to prove the statement.

If $f(a) \neq y_0 = 0$, then open, relative compact neighborhoods M_1 of a in M and W_1 of y_0 in W exist such that $f(\overline{M}_1) \cap \overline{W}_1 = \emptyset$. A constant $c > 0$ exists such that $|f(z)-y| \geq c$ for all $(z,y) \in M_1 \times W_1$. Hence the statement is trivially true.

Therefore, only the case $f(a) = y_0 = 0 \in W \cap U$ remains. Because $h_y^{-1}(x) = x + y$, open, relative compact neighborhoods U_1 and W_0 of $0 \in \mathbb{C}^n$ with $\overline{U}_1 \subset U$ and $\overline{W}_0 \subset W$ exist such that $h_y^{-1}(U_1) \subset U$ if $y \in W_0$. Hence $\overline{U}_1 \subset h_y(U)$ if $y \in \overline{W}_0$. Hence $\tilde{h}(U \times W_0)$ contains $U_1 \times W_0$. Therefore, $\tilde{h}^{-1}$ maps $U_1 \times W_0$ into $U \times W_0$. Hence $\tilde{\varphi} = (\tilde{h}^{-1})^*(\varphi)$ is a locally bounded form on $U_1 \times W_0$. Define $i_y\colon \mathbb{C}^n \to \mathbb{C}^n \times W$ by $i_y(z) = (z,y)$ if $z \in \mathbb{C}^n$. Then

$$i_y^*(\tilde{\varphi}) = i_y^*(\tilde{h}^{-1})^*(\varphi) = (h^{-1} \circ i_y)^*(\varphi) = (i_y \circ h_y^{-1})^*(\varphi)$$

$$= (h_y^{-1})^* i_y^*(\varphi) = (h^{-1})_y^*(\varphi_y).$$

Therefore $i_y^*(\tilde{\varphi})$ is measurable on U_1 for each $y \in W_0$. Observe, that $\tilde{h}(y,y) = (0,y)$. Hence, if φ if of class $C^\beta(\mu;0)$ on H, then $\tilde{\varphi}$ is of class $C^\beta(\mu;0)$ on $(U_1-\{0\}) \times W_0$.

Lemma A I.9 shall be applied. For the convenience of the reader, a translation table will now be given,

Here	W_1	e	W_0	U_1	V	n	h	h_y	$\tilde{h}$	m	q	M_0
AI.9	Y	0	Y_0	U	V	n	h	h_y	$\tilde{h}$	m	q	M_0

Here	M_1	F	$\|F\|$	k	r	K_1	$\tilde{\varphi}$	j_y	i_y	$\tilde{\lambda}\chi$	K_1'
AI.9	M	F	$\|F\|$	k	r	K	φ	j_y	i_y	χ	K'

Open, relative compact neighborhoods V and W_1 of $0 \in \mathbb{C}^n$ with $\overline{V} \subset U_1$ and $\overline{W}_1 \subset W_0$ exist such that $h_y(\overline{V}) \subset U_1$ for all $y \in \overline{V}$.

A biholomorphic map $\alpha: M_0 \to M_0'$ of an open neighborhood M_0 of a in M onto an open neighborhood M_0' of $0 = \alpha(a) \in \mathbb{C}^m$ exists such that $f(M_0) \subset U_1$. Define $M_1 = f^{-1}(V) \cap M_0$. Then M_1 is the desired neighborhood of a. Take any C^∞-function λ on M with compact support T in M_1. Define $K_1 = K \cap T$. Then K_1 is a compact subset of M_1. Define $\tilde{\lambda}: M_1 \times W \to \mathbb{R}$ by $\tilde{\lambda}(z,y) = \lambda(z)$ for $z \in M_1$ and $y \in W$.

For notational simplicity, identify $M_0 = M_0'$ such that α becomes the identity. Then $a = 0 \in \mathbb{C}^m$. Define $F: M_1 \times W_1 \to U_1 \times W_1$ by $F(z,y) = \tilde{h}(f(z),y) = (f(z)-y,y)$ if $z \in M_1$ and $y \in W_1$. Define $|F(z,y)| = |h(f(z),y)| = |f(z)-y|$. For $y \in W$, define $j_y: \mathbb{C}^m \to \mathbb{C}^m \times W$ by $j_y(z) = (z,y)$ if $z \in \mathbb{C}^m$. For $y \in W_1$ and $z \in M$,

$$(\tilde{h}^{-1} \circ F \circ j_y)(z) = \tilde{h}^{-1}(\tilde{h}(f(z),y)) = (f(z),y)$$

$$= i_y(f(z)) = (i_y \circ f)(z)$$

Hence $\tilde{h}^{-1} \circ F \circ j_y = i_y \circ f$ on M_1 if $y \in W_1$. Therefore

$$j_y^* \circ F^*(\tilde{\varphi}) = j_y^* \circ F^* \circ (\tilde{h}^{-1})^*(\varphi) = (\tilde{h}^{-1} \circ F \circ j_y)^*\varphi$$

$$= (i_y \circ f)^*(\varphi) = f^* \circ i_y^*(\varphi) = f^*(\varphi_y)$$

$$j_y^*(\tilde{\lambda}\chi) = \lambda\chi_y.$$

Then $\tilde{\lambda}\chi$ is locally bounded on $M_1 \times W_1$ and $j_y^*(\tilde{\lambda}\chi)$ is measurable for each $y \in W_1$. Define $K_1' = K_1 \times W_1 - F^{-1}(0,W_1)$. Then $K_1' = (K_1 \times W_1) \cap R$. Hence, if χ is of class $C^\beta(\mu;0)$ on R, then $\tilde{\lambda}\chi$ is of class $C^\beta(\mu;0)$ on K_1'.

Therefore Lemma AI.9 implies that $I_\lambda(y)$ exists for each $y \in W_1$ and that I_λ is locally bounded and measurable on W_1. If, in addition, φ is of class $C^\beta(\mu,0)$ on H and χ of class $C^\beta(\mu;0)$ on R, then I_λ is continuous on W_1. The Statement is proven; q.e.d.

Lemma AI.11.[33] Let M be a complex manifold of dimension m. Let K be a compact subset of M. Let $f: M \to \mathbb{C}^n$ be an open, holomorphic map. Let k,r be non-negative integers with $r \leq 2n$. Let φ be a measurable and locally bounded form of degree r on an open neighhood U of $f(K)$. Let χ be a measurable and locally bounded form of degree 2m - r on M. Then the integral

$$I = \int_K (\log \frac{1}{|f|})^k \frac{1}{|f|^r} f^*(\varphi) \wedge \chi$$

exists. Moreover, suppose that a constant $B > 0$ and a family of measurable functions h_ρ for $0 < \rho < 1$ is given such that $|h_\rho(z)| \leqq B$ for all $z \in K$ and all ρ with $0 < \rho < 1$. For $0 < \rho < 1$, define

$$L(\rho) = \{z \in K \mid \frac{\rho}{2} \leqq |f(z)| \leqq \rho\}$$

$$J(\rho) = \int_{L(\rho)} |h_\rho| (\log \frac{1}{|f|})^k \frac{1}{|f|^r} |f^*(\varphi) \wedge \chi|$$

Then $I(\rho) \to 0$ for $\rho \to 0$.

Proof. Theorem AI.9 implies the existence of I. For $0 < \rho < 1$, define

$$L'(\rho) = \{z \in K \mid |f(z)| \leqq \rho\}$$

$$0 < I(\rho) = \int_{L'(\rho)} (\log \frac{1}{|f|})^k \frac{1}{|f|^\rho} |f^*(\varphi) \wedge \chi| < \infty$$

Because $\bigcap_{0<\rho<1} L'(\rho) = K \cap f^{-1}(0)$ is a set of measure zero, $I(\rho) \to 0$ for $\rho \to 0$. Observe, that $I(\rho)$ and $J(\rho)$ have non-negative integrands. Hence, $0 \leqq J(\rho) \leqq BI(\rho)$, which implies $J(\rho) \to 0$ for $\rho \to \infty$, q.e.d.

<u>Definition AI.12.</u>[35] <u>Define $P = \{\rho \in \mathbb{R} | 0 < \rho < 1\}$. Then $g = \{g_\rho\}_{\rho \in P}$ is called a test family if the following conditions are satisfied:</u>

1. <u>Each function $g_\rho \cdot \mathbb{R} \to \mathbb{R}$ is of class C^∞ for $\rho \in P$.</u>
2. <u>If $\rho \in P$ and $x \in \mathbb{R}$ then $0 \leqq g_\rho(x) \leqq 1$.</u>
3. <u>If $\rho \in P$ and $x \leqq \frac{1}{2}\rho$, then $g_\rho(x) = 0$.</u>
4. <u>If $\rho \in P$ and $x \geqq \rho$, then $g_\rho(x) = 1$.</u>
5. <u>A constant $B > 0$ exists such that $\rho|g_\rho'(x)| \leqq B$ for all $x \in \mathbb{R}$ and $\rho \in I$.</u>

Obviously, 5. is equivalent to

5' <u>A constant $B' > 0$ exists such that $|xg_\rho'(x)| \leqq B'$ for all $x \in \mathbb{R}$ and $\rho \in P$.</u>

<u>Lemma AI.13.</u> <u>A test family exists.</u>

<u>Proof.</u> Take any C^∞-function h on $\mathbb{R}$ with $0 \leqq h \leqq 1$ such that $h(x) = 0$ for $x \leqq 0$ and $h(x) = 1$ for $x \geqq 1$. For $\rho \in P$ define g_ρ by

$$g_\rho(x) = h\left(\frac{2x-\rho}{\rho}\right) \qquad \text{if } x \in \mathbb{R}.$$

Conditions 1 to 4 are satisfied. Take B such that $2|h'(x)| \leqq B$ for $0 \leqq x \leqq 1$. This remains true for all $x \in \mathbb{R}$. Hence

$$\rho|g'_\rho(x)| = 2|h'(\frac{2x-\rho}{\rho})| \leqq B$$

if $x \in \mathbb{R}$ and $\rho \in P$, q.e.d.

Lemma AI.14. Let M be a complex manifold of dimension m. Let $f: M \to \mathbb{C}^n$ be an open, holomorphic map. Let K be a compact subset of M. Let φ be a measurable and locally bounded form of degree $r < 2n$ on an open neighborhood U of f(K). Define $t = \mathrm{Max}\,(r, 2n-2)$. Let χ be a measurable and locally bounded form of degree $2m - r - 1$ on M. Let $\{g_\rho\}_{\rho\in P}$ be a test family. For $\rho \in P$ define γ_ρ on M by $\gamma_\rho(z) = g_\rho(|f(z)|)$ for $z \in M$. Let k be a non-negative integer. Define

$$J_\rho = \int_K d\gamma_\rho \wedge (\log \frac{1}{|f|})^k \frac{1}{|f|^t} f^*(\varphi) \wedge \chi$$

Then $J_\rho \to 0$ for $\rho \to \infty$.

Proof. Define $L(\rho)$ as in Lemma AI.11. Define $h_\rho = |f|g'_\rho(|f|)$. A constant B exists such that $|h_\rho| \leqq B$ on M for all $\rho \in P$. Define

$$\eta = \frac{1}{2}\frac{1}{|z|} d|z|^2 \qquad \text{on } \mathbb{C}^n - \{0\}$$

and $\eta(0) = 0$. Then η is measurable and locally bounded on $\mathbb{C}^n$. Then

$$d\gamma_\rho = \frac{1}{|f|} h_\rho f^*(\eta)$$

If $t = r \leqq 2n - 2$, then

$$J_\rho = \int_{L(\rho)} h_\rho \left(\log \frac{1}{|f|}\right)^k \frac{1}{|f|^{r+1}} f^*(\eta \wedge \varphi) \wedge \chi$$

where $\eta \wedge \varphi$ has degree $r + 1 < 2n$. Lemma AI.11 implies $J_\rho \to 0$ for $\rho \to 0$.

If $r = 2n - 1$, then $t = 2n - 2$. On $M - f^{-1}(0)$, define $\tilde{\chi} = f^*(\eta) \wedge \chi$. On $f^{-1}(0)$, define $\tilde{\chi} = 0$. Then $\tilde{\chi}$ has locally bounded and measurable coefficients. Now

$$J_\rho = -\int_{L(\rho)} h_\rho \left(\log \frac{1}{|f|}\right)^k \frac{1}{|f|^r} f^*(\varphi) \wedge \tilde{\chi}.$$

Again, Lemma AI.11 implies $J_\rho \to 0$ for $\rho \to 0$, q.e.d.

Proposition AI.15. Let M be a complex manifold of dimension m. Let K be a compact subset of M. Let W and W' be open neighborhoods of $0 \in \mathbb{C}^n$. Let $h\colon W \to W'$ be a biholomorphic map with $h(0) = 0$. Let $f\colon M \to W$ be an open, holomorphic map. Let φ be a measurable and locally bounded form of degree $r < 2n$ on W. Let χ be a measurand locally bounded form of $2m - r - 1$ on M. Let $\{g_\rho\}_{\rho \in P}$ be a test family. Define γ_ρ on M by

$$\gamma_\rho(z) = g_\rho(|h(f(z))|)$$

for $z \in M$ if $\rho \in P$. Define $t = \operatorname{Min}(r, 2n-2)$. Let k be a non-negative integer. Define

$$J_\rho = \int_K d\gamma_\rho \wedge \left(\log \frac{1}{|f|}\right)^k \frac{1}{|f|^t} f^*(\varphi) \wedge \chi.$$

Then $J_\rho \to 0$ for $\rho \to \infty$.

Proof. Take ρ_0 with $0 < \rho_0 < 1$ such that $\{z \mid |z| \leqq \rho_0\} \subset W' \cap W$. A number ρ_1 with $0 < \rho_1 < \rho_0$ and positive numbers b_1, b_2 exist such that

$$b_1|z| \leqq |h(z)| \leqq b_2|z|$$

if $|z| \leqq \rho_1$. Define $F = h \circ f\colon M \to W'$. Then $\gamma_\rho = g_\rho(|F|)$. Define

$$K_1 = \{z \in K \mid |f(z)| \geqq \rho_1\}$$

Then K_1 is compact and $F(z) \neq 0$ if $z \leqq K_1$. Take positive numbers L, L_0 and L_1 such that $|f(z)| \leqq L$ and $|F(z)| \leqq L_0$ if $z \in K$ and $|F(z)| \geqq L_1$ if $z \in K_1$.

Define $K' = K - f^{-1}(0)$. If $z \in K'$ then $F(z) \neq 0$. Define $u\colon K' \to \mathbb{R}$ by

$$u(z) = \frac{|F(z)|}{|f(z)|} \qquad \text{for } z \in K'.$$

If $z \in K_1$, then $0 < L_1 L^{-1} \leqq U(z) \leqq L_0 \rho_1^{-1}$. If $z \in K' - K_1$, then $0 < b_1 \leqq u(z) \leqq b_2$. For $\mu = 0,1,\cdots,k$, define H_μ on M by $H_\mu(z) = 0$ if $z \in M - K'$ and by

$$H_\mu(z) = (\log u(z))^{k-\mu} u(z)^t.$$

Then H_μ is a bounded and measurable function on M. Moreover, $(h^{-1})^*(\varphi) = \tilde{\varphi}$ is a form of degree r on W' with measurable and locally bounded coefficients. Define

$$J_{\rho,\mu} = \int_K d\gamma_\rho \wedge \left(\log \frac{1}{|F|}\right)^\mu \frac{1}{|F|^t} F^*\varphi \wedge (H_\mu \chi)$$

Lemma AI.14 implies $J_{\rho,\mu} \to 0$ for $\rho \to 0$. On K'

$$\sum_{\mu=0}^{k} \binom{k}{\mu} H_\mu \left(\log \frac{1}{|F|}\right)^\mu \frac{1}{|F|^t}$$

$$= \sum_{\mu=0}^{k} \binom{k}{\mu} (\log u)^{k-\mu} \left(\log \frac{1}{|F|}\right)^\mu \left(\frac{u}{|F|}\right)^t$$

$$= \left(\log \frac{|F|}{|f|} + \log \frac{1}{|F|}\right)^k \frac{1}{|f|^t}$$

$$= \left(\log \frac{1}{|f|}\right)^k \frac{1}{|f|^t} .$$

Therefore,

$$J_\rho = \sum_{\mu=0}^{k} \binom{k}{\mu} J_{\rho\mu} \to 0 \qquad \text{for } \rho \to 0$$

q.e.d.

Recall, that $\upsilon(z) = \frac{1}{2} d^\perp d|z|^2$ and $\upsilon_p = \frac{1}{p!} \upsilon \wedge \cdots \wedge \upsilon$ p-times on $\mathbb{C}^n$. Let $z_1, \cdots, z_n$ be the coordinates on $\mathbb{C}^n$, then

$$\frac{1}{2}\,\partial|z|^2 \wedge \bar{\partial}|z|^2 \wedge \upsilon_{n-1}(z)$$

$$= \left(\frac{1}{2}\right)^n \sum_{\mu,\nu=1}^{n} \bar{z}_\mu z_\nu dz_\mu \wedge d\bar{z}_\nu \wedge \sum_{\rho=1}^{n} \prod_{\substack{\lambda=1\\ \lambda\neq\rho}}^{n} dz_\lambda \wedge d\bar{z}_\lambda$$

$$= \left(\frac{1}{2}\right)^n \sum_{\mu=1}^{n} |z_\mu|^2 dz_1 \wedge d\bar{z}_1 \wedge \cdots \wedge dz_n \wedge d\bar{z}_n$$

$$= |z|^2 \upsilon_n(z)\,.$$

If $n > 1$, then

$$\frac{1}{2}\, d^\perp d(|z|^{2-2n}\upsilon_{n-1}) = \frac{1}{2}\,\partial\bar{\partial}|z|^{2-2n} \wedge \upsilon_{n-1}$$

$$= \frac{1}{2}(1-n)|z|^{-2n}\partial\bar{\partial}|z|^2 \wedge \upsilon_{n-1}$$

$$+ \frac{1}{2}(n-1)n|z|^{-2n-2}\partial|z|^2 \wedge \bar{\partial}|z|^2 \wedge \upsilon_{n-1}$$

$$= (-(n-1)n\upsilon_n + (n-1)n\upsilon_n)|z|^{-2n} = 0$$

Also

$$+ \frac{1}{2}d^\perp|z|^{2-2n} = -\frac{1}{2}(n-1)|z|^{-2n}d^\perp|z|^2$$

$$= |z|^{2-2n}d^\perp \log \frac{1}{|z|}$$

Hence

$$d\left(d^{\perp} \log \frac{1}{|z|} \wedge \frac{\upsilon_{n-1}(z)}{|z|^{2n-2}}\right) = 0$$

on $\mathbb{C}^n - \{0\}$. The formula remains true for $n = 1$.

For $r > 0$, define $B_r = \{z \in \mathbb{C}^n \mid |z| < r\}$. Then $S_r = \overline{B}_r - B_r$ is a boundary manifold of S_r. The euclidean volume element σ_r on S_r is the unique C^∞-form of degree $2n - 1$ on the oriented manifold S_r which is invariant under the unitary group and which gives the volume

$$\int_{S_r} \sigma_r = \frac{2\pi^n}{(n-1)!} r^{2n-1}.$$

Because σ_r is invariant and the volume is positive, $\sigma_r > 0$.

Lemma AI.16.[36] For $r > 0$, let $j_r: S_r \to \mathbb{C}^n$ be the inclusion map. On $\mathbb{C}^n - \{0\}$, define φ by

$$\varphi(z) = d^{\perp} \log \frac{1}{|z|} \wedge \upsilon_{n-1}(z);$$

then $\sigma_r = r j_r^*(\upsilon)$ on S_r.

Proof. Obviously, $j_r^*(\varphi)$ is invariant under the unitary group. Define $f: \mathbb{C}^n - \{0\} \to \mathbb{R}$ by $f(z) = |z|$. Then f is a regular map. As sets $f^{-1}(r) = S_r$ for each r. But as boundary manifold S_r carries the opposite orientation to S_r as fiber manifold of f (See Appendix AII Lemma AII3.3). Therefore, Appendix AII Theorem AII4.11 and Lemma AII4.6 imply

$$J = \int_{\mathbb{C}^n} e^{-|z|^2} d|z|^2 \wedge \varphi(z) = \int_0^\infty e^{-r^2} \int_{S_r} j_r^*(\varphi) dr^2 .$$

Define $m_r : S_1 \to S_r$ by $m_r(z) = rz$. Then m_r is an orientation preserving diffeomorphism. Obviously,

$$m_r^* j_r^*(\varphi) = r^{2n-2} j_1^*(\varphi)$$

on S_1. Hence

$$\int_{S_r} j_r^*(\varphi) = r^{2n-2} \int_{S_1} j_1^*(\varphi) .$$

Therefore

$$J = \int_0^\infty e^{-r^2} r^{2n-2} dr^2 \int_{S_1} j_1^*(\varphi) = (n-1)! \int_{S_1} j_1^*(\varphi).$$

Now

$$d|z|^2 \wedge \varphi(z) = \tfrac{1}{2}(\partial|z|^2 + \bar\partial|z|^2) \wedge (\bar\partial|z|^2 - \partial|z|^2)|z|^{-2} \upsilon_{n-1}(z)$$

$$= i|z|^{-2} \partial|z|^2 \wedge \bar\partial|z|^2 \wedge \upsilon_{n-1}(z)$$

$$= 2\upsilon_n(z)$$

which implies

$$J = 2\int_{\mathbb{C}^n} e^{-|z|^2} \upsilon_n(z) = 2\Big(\int_{\mathbb{C}} e^{-|z|^2} \upsilon(z)\Big)^n$$

$$= 2\Big(\int_0^\infty \int_0^{2\pi} e^{-t^2} t\,dt\,d\varphi\Big)^n = 2\pi^n .$$

Hence

$$\int_{S_r} j_r^*(\varphi) = r^{2n-2} \int_{S_1} j_1^*(\varphi) = \frac{2\pi^n}{(n-1)!}\, r^{2n-2} ,$$

which implies $\sigma_r = r j_r^*(\varphi)$; q.e.d.

Lemma AI.17. Let B be any open and bounded neighborhood of $0 \in \mathbb{C}^n$. Suppose that $S = \overline{B} - B$ is a boundary manifold of B. Then

$$\int_S d^\perp \log \frac{1}{|z|} \wedge \frac{\upsilon_{n-1}(z)}{|z|^{2n-2}} = \frac{\pi^{n-1}}{(n-1)!} .$$

Proof. Take $r > 0$ such that $\overline{B}_r \subset B$. Stokes Theorem implies

$$\int_S d^\perp \log \frac{1}{|z|} \wedge \frac{\upsilon_{n-1}(z)}{|z|^{2n-2}} = \int_{S_r} \frac{1}{|z|^{2n-2}}\, \varphi = \frac{2\pi^n}{(n-1)!}$$

q.e.d.

Theorem AI.18.[37] Let W and W' be open neighborhoods of $0 \in \mathbb{C}^n$. Let $h: W \to W'$ be a biholomorphic map with $h(0) = 0$. Take ρ_0 such that $0 < \rho_0 < 1$ and $\overline{B}_{\rho_0} \subset W'$. For $0 < \rho < \rho_0$, define $B_\rho(h) = h^{-1}(B_\rho)$ and $S_\rho(h) = h^{-1}(S_\rho)$. Then $\overline{B}_\rho(h)$ is compact and contained in W. Moreover, $S_\rho(h) = \overline{B}_\rho(h) - B_\rho(h)$ is a boundary manifold of $B_\rho(h)$.

Let $\{g_\rho\}_{\rho\in P}$ be a test family. For $0 < \rho \leqq \rho_0$, define $U_\rho = g_\rho(|h|)$ on W.

Let M be a complex manifold of dimension $m \geqq n$. Define $q = m - n$. Let χ be a continuous form of bidegree (q,q) on M. Let $H \neq \emptyset$ be an open and relative compact subset of M. Let T be the support of χ on $\overline{H} - H$.

Let $f: M \to W$ be an open holomorphic map. Denote by $\nu_f(z)$ the multiplicity of f at $z \in M$. Suppose that $f^{-1}(0) \cap T$ is a set of measure zero on the complex space $f^{-1}(0)$ if $q > 0$ and that $f^{-1}(0) \cap T = \emptyset$ if $q = 0$. For $0 < \rho \leqq \rho_0$, define $\gamma_\rho: M \to \mathbb{R}$ by $\gamma_\rho(z) = u_\rho(f(z))$. Define $F = f^{-1}(0) \cap H$. For $0 < \rho \leqq \rho_0$, define

$$J_\rho = \int_H d\gamma_\rho \wedge d^\perp \log \frac{1}{|f|} \wedge \frac{f^*(\upsilon_{n-1})}{|f|^{2n-2}} \wedge \chi .$$

Then[38)]

$$J_\rho \to \frac{2\pi^n}{(n-1)!} \int_F \nu_f \chi ,$$

Proof. Because $\overline{H} \cap \operatorname{supp}(d\gamma_\rho)$ is a compact subset of $M - f^{-1}(0)$, the integral J_ρ exists. For $w \in W$, define $F(w) = H \cap f^{-1}(w)$ and

$$J(w) = \int_{F(w)} \nu_f \chi .$$

By [27] Theorem 3.9 if $q > 0$, and by [26] Proposition 3.2 if $q = 0$ the integral, respectively sum, J is continuous at $0 \in W$. On $W - \{0\}$, define ψ and φ by

$$\psi(w) = d^{\perp} \log \frac{1}{|w|} \wedge \frac{\upsilon_{n-1}(w)}{|w|^{2n-2}} = |w|^{2-2n} \varphi(w)$$

[27] Proposition 2.9 if $q > 0$ resp. [27] Proposition 2.8 if $q = 0$ implies

$$J_\rho = \int_W J \, du_\rho \wedge \psi .$$

Define

$$I_\rho^0 = \int (J(w) - J(0)) \, du_\rho \wedge \psi$$

$$I_\rho^1 = \int_W du_\rho \wedge \psi .$$

Then $J_\rho = I_\rho^0 + J(0) \cdot I_\rho^1$ for $0 < \rho \leqq \rho_0$. If $w \in B_{\frac{\rho}{2}}(h)$ then $h(w) \in B_{\frac{\rho}{2}}$ and $g_\varrho(|h(w)|) = 0$. If $w \in W - B_\varrho(h)$, then $h(w) \in W' - B_\varrho$ and $g_\varrho(|h(w)|) = 1$. Observe, that $d\psi = 0$. Stoke's theorem and Lemma AI.17 imply

$$I_\rho^1 = \int_{B_\rho(h)} d(u_\rho \wedge \psi) = \int_{S_\rho(h)} \psi = \frac{2\pi^n}{(n-1)!} .$$

Therefore, it only remains to be shown that $I_\rho^0 \to 0$ for $\rho \to 0$.

Now, the matrix notation is adopted. Let A^t be the transposed matrix of A. If $w \in \mathbb{C}^n$, then $w = (w_1, \cdots, w_n)$ is considered to be a matrix of 1 line and n columns. Hence w^t is defined. Take ρ_1 with $0 < \rho_1 < \rho_0$ such that $\overline{B}_{\rho_1} \subset W$. A holomorphic matrix function A exists in a neighborhood of $\overline{B}_{\rho_1}$ and a non-singular, constant matrix D (the Jacobian at 0) such that

$$h(w) = wD + wA(w)w^t$$

in a neighborhood of $\overline{B}_{\rho_1}$. Constants $c_2 > c_1 > 0$ exist such that

$$c_1|w| \leqq |wD| \leqq c_2|w|$$

for all $w \in \mathbb{C}^n$. A constant $c_3 > 0$ exists such that

$$|wA(w)w^t| \leqq c_3|w|^2$$

for all $w \in \overline{B}_{\rho_1}$. Define $c_4 = c_2 + \rho_1 c_3$. Then

$$|h(w)| \leqq (c_2 + |w|c_3)|w| \leqq c_4|w|$$

for $w \in \overline{B}_{\rho_1}$. Take ρ_2 with $0 < \rho_2 < \rho_1$ such that $c_5 = c_1 - c_3\rho_2 > 0$. For $w \in \overline{B}_{\rho_2}$:

$$|h(w)| \geqq |wD| - |wA(w)w^t|$$

$$\geqq (c_1 - c_3|w|)|w| \geqq c_5|w| .$$

Therefore

$$c_5|w| \leqq |h(w)| \leqq c_4|w| .$$

for $w \in \overline{B}_{\rho_2}$. Because $\overline{B}_{\rho_0}(h) - B_{\rho_2}$ is compact, constants c_6 and c_7 with $0 < c_6 \leqq c_5 \leqq c_4 \leqq c_7$ exist such that

$$c_6|w| \leqq |h(w)| \leqq c_7|w|$$

for $w \in \overline{B}_{\rho_0}(h) - B_{\rho_2}$, hence for all $w \in \overline{B}_{\rho_0}(h) \cup \overline{B}_{\rho_2}$.

Define $b = (c_6)^{-1}$ and $a = \frac{1}{2}(c_7)^{-1}$. Take ρ_3 with $0 < \rho_3 < \rho_2$ such that $\rho_3 b < \rho_2$. Because $a \leqq b$, also $\rho_3 a < \rho_2$. Take any ρ with $0 < \rho \leqq \rho_3$. If $|w| \leqq a\rho$ then $w \in B_{\rho_2}$ and $|h(w)| \leqq c_7|w| \leqq \frac{1}{2}$. Therefore, $u_\rho(w) = g_\rho(|h(w)|) = 0$. If $w \in W$ and $|w| \geqq b_\rho$, then $u_\rho(w) = 1$. For, assume this would be wrong. Then $|h(w)| < \rho \leqq \rho_0$. Hence $w \in B_{\rho_0}(h)$. Therefore

$$|w| \leqq b|h(w)| < b_\rho \leqq |w|$$

Contradiction! Consequently, du_ρ has its support in $\overline{B}_{b_\rho} - B_{a_\rho}$. Define $K = \overline{B}_b - B_a$. Then

$$I_\rho^0 = \int_{w \in K} (J(\rho w) - J(0))d(u_\rho(\rho w)) \wedge \psi(w)$$

because $\psi(\rho w) = \psi(w)$.

If $w \in \overline{B}_b - B_a$ and $0 < \rho < \rho_3$, then $\rho w \in \overline{B}_{\rho_1}$. Hence

$$\frac{1}{\rho}d(h(\rho w)) = dwD + \rho(dw)A(\rho w)w^t + \rho wA(\rho w)dw^t$$

$$+ \rho^2 w(dA)(\rho w)w.$$

Hence this differential has uniformly bounded coefficients on K for $0 < \rho \leqq \rho_3$. Because $\rho g_\rho'(|h(\rho w)|)$ is also uniformly bounded, the same is true for

$$d(u_\rho(\rho w))$$
$$= \rho g_\rho'(|h(\rho w)|)[(\frac{d(h(\rho w))}{\rho}|\frac{h(\rho w)}{|h(\rho w)|}) + (\frac{h(\rho w)}{|h(\rho w)|}|\frac{d(h(\rho w))}{\rho})]$$

For $0 < \rho \leqq \rho_3$, define Δ_ρ on K by

$$\Delta_\rho(w)\upsilon_n(w) = d(u_\rho(\rho w)) \wedge \psi(w).$$

A constant $c > 0$ exists such that $|\Delta_\rho(w)| < c$ for all $w \in K$ and all ρ with $0 < \rho \leqq \rho_3$. Because J is continuous on 0,

$$(J(\rho w) - J(0))\Delta_\rho(w) \to 0 \qquad \text{for } \rho \to 0$$

uniformly on K. Therefore

$$I_\rho^0 = \int_{w \in K} (J(\rho w) - J(0))\Delta_\rho(w)\upsilon_n(w) \to 0$$

for $\rho \to 0$; q.e.d.

Appendix II

The Fiber Integral.

The integration of a differential form ψ of degree p over the fibers of a regular map with fiber dimension $q \leqq p$ is an important operator. Although, this operator is used in various papers and the operator seems to be known widely, no account seems to have appeared, which gives complete statements and precise proofs. Therefore, an attempt is made, to give such an account here.

The topic can be presented in different ways. The operators have smooth and elegant formulation, if the maps are proper and regular. However, this may be restrictive in some application, therefore a more general approach will be considered here. The theory can be developed from a differential geometric point of view, or by the use of distributions. Here, the first approach has been adopted, so that the concept of distributions does not have to be introduced or presupposed here. Also, most of the differential geometric concepts are rather elementary and have to be used anyway.

§1 Integration of vector valued forms

Let M be a manifold of dimension m. If not otherwise stated, "manifold" means a pure dimensional, paracompact, oriented, differentiable manifold of class C^∞. Let $\pi: E \to M$ be a differentiable fiber bundle over M. For $U \subseteq M$, let $\Gamma(U,E)$ be the set of sections in E over U. Let $\Gamma_k(U,E)$ be the set of sections of class C^k in E over U, however, "class C^k" may be defined. (Besides the usual conventions, $k = \mu$ means "measurable" on a measurable set U, and $k = \lambda$ means "locally integrable".)

The complexified cotangent bundle of M is denoted by $T = T(M)$. Its p-fold, exterior product is $T^p = T^p(M) = \Lambda_p T$, whose sections are the forms of degree p on M. Let D_A be the set of those measurable forms $\omega \in \Gamma_\mu(A,T^m)$ of bidegree m, which are integrable over the measurable subset A of M. For $\omega \in D_A$, the integral $\int_A \omega$ is defined and its properties are assumed to be known.[39)]

Let V be a complex vector space of dimension n. Then $V_M = V \times M$ is the trivial bundle over M with general fiber V. An element $v \in V$ defines a global section in V_M, called a <u>"constant" section</u> and again denoted by v, namely $v(x) = (v,x)$ for all $x \in M$. If $(e_1,\cdots,e_n) = e$ is a base of V, then $\omega \in \Gamma_\mu(A,T^m \otimes V_M)$ is given by

(1.1) $$\omega = \sum_{\nu=1}^{n} \omega_\nu \otimes e_\mu$$

where the coefficients $\omega_\nu \in \Gamma_\mu(A,T^m)$ are unique. Now, ω is said to be <u>integrable over A</u>, if and only if $\omega_\nu \in D_A$ for $\nu = 1,\cdots,n$ and if so, the integral of ω over A is defined by

(1.2) $$\int_A \omega = \sum_{\nu=1}^{n} e_\nu \int_A \omega_\nu$$

Both, integrability and integral are independent of the choice of the base e. Let $D_A(V)$ be the set of $\omega \in \Gamma_\mu(A, T^m \otimes V_M)$, which are integrable over A. Obviously, $D_A(V)$ is a complex vector space over $\mathbb{C}$, and a module over the ring of bounded, measurable functions on A. The integral is a linear map of $D_A(V)$ into $\mathbb{C}$.

The following statements are easily proven, and the proofs are left to the reader.[40)]

<u>Lemma A II 1.1: If $\omega_\nu \in D_A$ and $v_\nu \in V$ for $\nu = 1,\cdots,s$, then $\omega = \sum_{\nu=1}^{s} \omega_\nu \otimes v_\nu \in D_A(V)$ with</u>

(1.3) $$\int_A \omega = \sum_{\nu=1}^{s} v_\nu \int_A \omega_\nu .$$

<u>Lemma A II 1.2. If A has measure zero, then $\int_A \omega = 0$.</u>

<u>Lemma A II 1.3. If $\omega \in \Gamma(A, T^m \otimes V_M)$, if $A = A_1 \cup A_2$, if A_1 and A_2 are measurable, and if $\omega \in D_{A_\nu}(V)$ for $\nu = 1,2$, then $\omega \in D_A(V)$ and $\omega \in D_{A_1 \cap A_2}(V)$ with</u>

(1.4) $$\int_A \omega = \int_{A_1} \omega + \int_{A_2} \omega - \int_{A_1 \cap A_2} \omega$$

Lemma A II 1.4. If $\omega \in D_A(V)$, if $B \subseteq A$, and if B is measurable, then $\omega \in D_B(V)$.

Lemma A II 1.5. If $\omega \in D_A(V)$, if $A = \bigcup_{\nu=1}^{\infty} A_\nu$, if each A_ν is measurable, and if $A_\nu \cap A_\mu$ has measure zero for $\nu \neq \mu$, then

$$(1.5) \qquad \int_A \omega = \sum_{\nu=1}^{\infty} \int_{A_\nu} \omega.$$

If V and W are complex vector spaces of dimension n and r respectively, and if $\alpha: V \to W$ is linear, then α extends to $\alpha: V_M \to W_M$ by $\alpha(v,x) = (\alpha(v),x)$ and to

$$(1.6) \qquad \alpha = \mathrm{Id} \otimes \alpha: T^m \otimes V_M \to T^m \otimes W_M$$

which induces a map α on the sections:

$$(1.7) \qquad \alpha: \Gamma_\mu(A, T^m \otimes V_M) \to \Gamma_\mu(A, T^m \otimes W_M)$$

Lemma A II 1.6. The map α of (1.7) restricts to $\alpha: D_A(V) \to D_A(W)$. If $\omega \in D_A(V)$ is given by (1.1), then

$$(1.8) \qquad \int_A \alpha(\omega) = \sum_{\nu=1}^{n} \alpha(e_\nu) \int_A \omega_\nu = \alpha(\int_A \omega)$$

Therefore, the following diagram commutes

$$(1.9)\qquad \begin{array}{ccc} D_A(V) & \xrightarrow{\ \alpha\ } & D_A(W) \\ \Big\downarrow \int_A & & \Big\downarrow \int_A \\ V & \xrightarrow{\ \alpha\ } & W \end{array}$$

Let $f\colon M \to N$ be a diffeomorphism onto a manifold N. Here, "diffeomorphism" means an orientation preserving diffeomorphism of class C^∞, unless stated otherwise. Let A be a measurable subset of N. Then $B = f^{-1}(A)$ is measurable in M. Each form $\omega \in D_A$ pulls back to a form $f^*(\omega) \in D_B$ with $\int_A \omega = \int_B f^*(\omega)$.

<u>Lemma A II 1.7.</u> <u>If $\omega \in D_A(V)$ is represented by (1.1), then $f^*\colon D_A(V) \to D_B(V)$ is well-defined by $f^*(\omega) = \sum_{\nu=1}^{n} f^*(\omega_\nu) \otimes e_\nu$ where $f^*(\omega)$ does not depend on the base e. Moreover</u>

$$(1.10)\qquad \int_A \omega = \int_B f^*(\omega) \qquad \underline{\text{with}}\ B = f^{-1}(A).$$

Let $\alpha\colon U \to U'$ be a diffeomorphism of an open subset U of M onto an open subset U' of $\mathbb{R}^m$ with $\alpha = (x_1, \cdots, x_m)$. Define $\underset{p}{\otimes} T^m = T^m \otimes \cdots \otimes T^m$ (p-times). If $\omega \in \Gamma(A, \underset{p}{\otimes} T^m)$, then

$$(1.11)\qquad \omega = \omega_0 \underset{p}{\otimes} dx_1 \wedge \cdots \wedge dx_m$$

where ω_0 is a function on $A \cap U$. If $x \in U \cap A$, then $\omega(x) \geqq 0$ (resp. $\omega(x) > 0$) if and only if $\omega_0(x) \geqq 0$ (resp. $\omega_0(x) > 0$). Because M is orineted, this definition does not depend on α. Write $\omega \geqq \psi$ (resp. $\omega > \psi$) if and only if $\omega(x) - \psi(x) \geqq 0$ (resp. $\omega(x) - \psi(x) > 0$) for all $x \in A$. This defines a partial ordering on $\Gamma(A, \underset{p}{\otimes} T^m)$.

Let $p = 2q$ be even. If $0 \leqq \omega \in \Gamma(A, \underset{2q}{\otimes} T^m)$ is represented as by (1.11), then $\sqrt{\omega} \in \Gamma(A, \underset{q}{\otimes} T^m)$ is well-defined by

$$(1.12) \qquad \sqrt{\omega} = \sqrt{\omega_0} \underset{q}{\otimes} dx_1 \wedge \cdots \wedge dx_m$$

which is independent of the coordinate system α, because M is oriented. Of course $\sqrt{\omega} \geqq 0$.

If $\omega \in \Gamma(A, \underset{p}{\otimes} T^m)$, then $\omega \otimes \overline{\omega} \in \Gamma(A, \underset{2p}{\otimes} T^m)$ with $\omega \otimes \overline{\omega} \geqq 0$. Define $|\omega| = \sqrt{\omega \otimes \overline{\omega}} \in \Gamma(A, \underset{p}{\otimes} T^m)$. Then $|\omega| \geqq 0$ (resp. $|\omega| > 0$ if $\omega(x) \neq 0$ for all $x \in M$). Moreover, if $\omega \in \Gamma(A, \underset{p}{\otimes} T^m)$ and $\psi \in \Gamma(A, \underset{q}{\otimes} T^m)$, then

$$(1.13) \qquad |\omega \otimes \psi| = |\omega| \otimes |\psi|$$

If $\omega \in \Gamma(A, \underset{p}{\otimes} T^m)$ and $\psi \in \Gamma(A, \underset{p}{\otimes} T^m)$ then

$$(1.14) \qquad |\omega + \psi| \leqq |\omega| + |\psi|$$

Let $(\,|\,): V \times V \to V$ be a positive definite hermitian form on V, called a <u>hermitian product</u>. The associated norm is defined by $|v| = \sqrt{(v|v)}$. This form extends to a positive definite hermitian form on $\Gamma(A, T^m \otimes V_M)$ with values in $\Gamma(A, T^m \otimes T^m)$: If $e = (e_1, \ldots, e_n)$ is a base of V, if ω and ψ are elements of $\Gamma(A, T^m \otimes V_M)$ with representations as in (1.1), then

$$(1.15) \qquad (\omega|\psi) = \sum_{\mu,\nu=1}^{n} (e_\mu|e_\nu)\omega_\mu \otimes \overline{\psi}_\nu$$

is well-defined independently of the base e. Obviously, $(\omega|\psi)$ is linear in ω with $(\omega|\psi) = \overline{(\psi|\omega)}$ and with $(\omega|\omega) \geqq 0$ where $(\omega|\omega) > 0$ if $\omega(x) \neq 0$ for all $x \in A$. Define the norm $|\omega|$ of ω by $|\omega| = \sqrt{(\omega|\omega)} \in \Gamma(A, T^m)$. The use of an orthonormal base e and local coordinates $\alpha\colon U \to U'$ implies immediately

$$(1.16) \qquad |(\omega|\psi)| \leqq |\omega||\psi|$$

$$(1.17) \qquad |\omega + \psi| \leqq |\omega| + |\psi|$$

if ω and ψ belong to $\Gamma(A, T^m \otimes V_M)$.

<u>Lemma A II 1.8.</u>[41)]

a) <u>If $\omega \in \Gamma_\mu(A, T^m \otimes V_M)$, then $\omega \in D_A(V)$ if and only if $|\omega| \in D_A$.</u>

b) <u>If $\omega \in D_A(V)$, if $B \subseteq A$, and if B is measurable, then</u>

$$|\int_B \omega| \leqq \int_A |\omega| .$$

c) <u>If $\omega \in \Gamma_\mu(A, T^m \otimes V_M)$, if $\psi \in D_A$ with $|\omega| \leqq \psi$, then $\omega \in D_A(V)$.</u>

d) <u>If $\omega_\nu \in D_A(V)$ for all $\nu \in \mathbb{N}$, if $\omega_\nu \to \omega$ for $\nu \to \infty$ and if $|\omega_\nu| \leqq \psi \in D_A$ then $\omega \in D_A(V)$ and $\int_A \omega_\nu \to \int_A \omega$ for $\nu \to \infty$. (Lebesgue).</u>

<u>Proof.</u> a) Let $e = (e_1, \cdots, e_n)$ be an orthonormal base of V. If ω if represented by (1.1), then $\omega \in D_A(V)$ if and only if $\omega_\nu \in D_A$ for $\nu = 1, \cdots, n$, which is true if and only if $|\omega_\nu| \in D_A$ for $\nu = 1, \cdots, n$. Now

$$|\omega_\nu| \leqq |\omega| \leqq |\omega_1| + \cdots + |\omega_n|$$

implies $|\omega| \in D_A$ if and only if $|\omega_\nu| \in D_A$ for $\nu = 1, \cdots, n$. This proves a).

b) Define $a = \int_B \omega \in V$. If $a = 0$, statement b) is trivially true. Suppose $a = \sum_{\nu=1}^n a_\nu e_\nu \neq 0$. Then $|a| \neq 0$ and

$$|a|^2 = |\sum_{\nu=1} a_\nu \int_B \overline{\omega}_\nu| = |\int_B \sum_{\nu=1}^n a_\nu \overline{\omega}_\nu|$$

$$\leqq \int_B |\sum_{\nu=1} a_\nu \overline{\omega}_\nu| \leqq \int_B |a||\omega| \leqq |a| \int_A |\omega|$$

$$|\int_B \omega| = |a| \leqq \int_A |\omega| .$$

c) If $\omega = \sum_{\nu=1}^n \omega_\nu \otimes e_\nu$, then $|\omega_\nu| \leqq |\omega| \leqq \psi$; hence $\omega_\nu \in D_A$ for $\nu = 1, \cdots, n$, which implies $\omega \in D_A(V)$.

d) If $\nu \in \mathbb{N}$, then $\omega_\nu = \sum_{\delta=1}^n \omega_{\delta\nu} \otimes e_\delta$ and $\omega_{\delta\nu} \to \omega_\delta$ for $\nu \to \infty$ if

$s \leqq \delta \leqq n$, where $\omega = \sum_{\delta=1}^{n} \omega_\delta \otimes e_\delta$. Because $|\omega_{\delta\nu}| \leqq |\omega_\nu| \leqq \psi \in D_A$

$$\int_A \omega_\nu = \sum_{\delta=1}^{n} e_\delta \int_A \omega_{\delta\nu} \to \sum_{\delta=1}^{n} e_\delta \int_A \omega_\delta = \int_A \omega$$

for $\nu \to \infty$; q.e.d.

Let A and B vector spaces or vector bundles. Define $B^p = B \wedge \ldots \wedge B$ (p-times). A product

(1.18) $$\Delta\colon (A \otimes B^p) \times B^q \to A \otimes B^{p+q}$$

shall be defined. Take $\omega \in A \otimes B^p$ and $\psi \in B^q$. Let $(e_1, \cdots, e_r) = e$ be a base of B^p (respectively a local frame field). Then

(1.19) $$\omega = \sum_{\mu=1}^{r} \omega_\mu \otimes e_\mu$$

with $\omega_\mu \in A$. Define

(1.20) $$\omega \Delta \psi = \sum_{\mu=1}^{r} \omega_\mu \otimes (e_\mu \wedge \psi)$$

Obviously, the definition independent of the choice of the base e and the product Δ is bilinear.

If A and B are vector bundles over M, and if U is a subset of M, then Δ extends to sections

(1.21) $$\Delta\colon \Gamma(U, A \otimes B^p) \times \Gamma(U, B^q) \to \Gamma(U, A \otimes B^{p+q})$$

by $(\omega \Delta \psi)(x) = \omega(x) \Delta \psi(x)$ for each $x \in M$.

Lemma A II 1.9. Let A be a measurable subset of M. Take $v \in V^q$ and regard v as a constant section in V^q_M. Take $\omega \in D_A(V^p) \subseteq \Gamma(A, T^m \otimes V^p_M)$. Then $\omega \Delta v \in D_A(V^{p+q})$ and

$$\int_A \omega \Delta v = (\int_A \omega) \wedge v \in V^{p+q}$$

Proof. Take a base $e = (e_1, \cdots, e_r)$ of V^p. Then $\omega = \sum_{\mu=1}^{r} \omega_\mu \otimes e_\mu$ with $\omega_\mu \in D_A$ for $\mu = 1, \cdots, r$ and

$$\omega \Delta v = \sum_{\mu=1}^{r} \omega_\mu \otimes (e_\mu \wedge v) \in D_A(V^{p+q})$$

$$\int_A \omega \Delta v = \sum_{\mu=1}^{r} (\int_A \omega_\mu) e_\mu \wedge v = (\int_A \omega) \wedge v$$

by Lemma A II 1.1, q.e.d.

Let E be a vector bundle over M. Let $\alpha: E \to E$ be an additive isomorphism such that $\alpha(cv) = \bar{c}\alpha(v)$ if $v \in E$ and $c \in \mathbb{C}$, and such that $\alpha \circ \alpha$ is the identity. Then α is called a conjugation of E. For $v \in E$, define $\bar{v} = \alpha(v)$. Then $v \in E$ is said to be real if and only if $\bar{v} = v$. For $v \in E$, define

$$\operatorname{Re} v = \frac{1}{2}(v+\bar{v}) \qquad \operatorname{Im} v = \frac{1}{2i}(v-\bar{v})$$

as the real and imaginary part of v. Both are real and

$v = \text{Re } v + i \text{ Im } v$. Moreover, if $v = u + iw$ where u and w are real then $u = \text{Re } v$ and $w = \text{Im } v$. Then the real elements of E form a subbundle E_R of E, which is a real vector bundle, such that $E = E_R \oplus iE_R$. On the exterior product E^p, a conjugation is uniquely defined by

$$\overline{v_1 \wedge \cdots \wedge v_r} = \overline{v}_1 \wedge \cdots \wedge \overline{v}_p$$

where $v_\mu \in E_x$ for $\mu = 1,\ldots,p$. If $E_1,\ldots,E_p$ are complex vector bundles with conjugation, then a conjugation is defined on $E_1 \otimes \ldots \otimes E_p$ by

$$v_1 \otimes \ldots \otimes v_p = \overline{v}_1 \otimes \ldots \otimes \overline{v}_p$$

The conjugation on E clearly extends to a conjugation of the sections s in E by $\overline{s}(x) = \overline{s(x)}$. Because vector spaces are vector bundles over a point, the concept of conjugation is extended to complex vector spaces.

Let $T = T(M)$ be the complexified cotangent bundle of the manifold M. Let $a\colon U \to U'$ be a diffeomorphism of an open subset U of M onto an open subset U' of $\mathbb{R}^m$. Set $\alpha = (x_1,\ldots,x_m)$. Then $dx_1,\ldots,dx_m$ is a frame field of T(M) over U. If $\omega \in T_x(M)$ with $x \in U$ then

$$\omega = \sum_{\mu=1}^{m} \omega_\mu \, dx_\mu$$

on U, where $\omega_\mu \in \mathbb{C}$. A conjugation is defined by

$$\overline{\omega} = \sum_{\mu=1}^{m} \overline{\omega}_\mu \, dx_\mu$$

independent of the choice of α. Hence if V is a vector space with conjugation, then V_M and T^m are vector bundles with conjugation. Hence $T^m \otimes V_M$ is a vector bundle with conjugation.

Lemma A II 1.10. Let V be a vector space with conjugation. Let A be a measurable subset of M. Then $\omega \in D_A(V)$ if and only if $\overline{\omega} \in D_A(V)$ and $\overline{(\int_A \omega)} = \int_A \overline{\omega}$.

Proof. If ω is represented by (1.1), then $\overline{\omega} = \sum_{\nu=1}^{n} \overline{\omega}_\nu \otimes \overline{e}_\nu$ where $\omega_\nu \in D_A$ if and only if $\overline{\omega}_\nu \in D_A$. If so, then

$$\overline{\int_A \omega} = \sum_{\nu=1}^{m} \overline{e}_\nu \overline{\int_A \omega_\nu} = \sum_{\nu=1}^{m} \overline{e}_\nu \int_A \overline{\omega}_\nu = \int_A \overline{\omega} ,$$

q.e.d.

§2 A remark on short exact sequences.

As before, define $\Delta_p = \{x \in N \mid 1 \leqq x \leqq p\}$ for $p \in N$. For integers p and q with $0 < p \leqq q$, let $T(p,q)$ be the set of all increasing, injective maps $\mu: \Delta_p \to \Delta_q$. If V is any vector space, if $a_1, \ldots, a_q$ are vectors in V, and if $\mu \in T(p,q)$, define

(2.1) $$a_\mu = a_{\mu(1)} \wedge \cdots \wedge a_{\mu(p)}.$$

Let M be a manifold of dimension m. Let

(2.2) $$0 \longrightarrow B \xrightarrow{\beta} E \xrightarrow{\alpha} A \longrightarrow 0$$

be an exact sequence of differentiable vector bundles over M with fiber dimension n of E, s of B and q = n - s of A. An exact sequence

(2.3) $$0 \longrightarrow A \xrightarrow{\gamma} E \xrightarrow{\delta} B \longrightarrow 0$$

is said to be a <u>splitting</u> of (2.2) if and only if

(2.4 $$\alpha \circ \gamma = \mathrm{Id}, \quad \delta \circ \beta = \mathrm{Id}, \quad \gamma \circ \alpha + \beta \circ \delta = \mathrm{Id}.$$

are the identity maps. Such splittings exist and can be obtained with the use of a hermitian metric along the fibers of E. Suppose that such a splitting is given. Then $\gamma(A)$ and $\beta(B)$ are subbundles of E with

(2.5) $$E = \gamma(A) \oplus \beta(B).$$

Let $\gamma(A)^a \wedge \gamma(B)^b$ be the subbundle of E^{a+b} generated by $\varphi \wedge \psi$ with $\varphi \in \gamma(A)^a_x$ and $\psi \in \beta(B)^b_x$ with $x \in M$. Then

(2.6) $$\sigma\colon \gamma(A)^a \wedge \beta(B)^b \to \gamma(A)^a \otimes \beta(b)^b$$

is the standard isomorphism defined by $\sigma(\varphi \wedge \psi) = \varphi \otimes \psi$. Observe that β and γ define isomorphisms

(2.7) $$\beta\colon B^b \to \beta(B)^b \qquad \gamma\colon A^a \to \gamma(A)^a .$$

Therefore, an isomorphism

(2.8) $$\eta = \gamma^{-1} \otimes \beta^{-1} : \gamma(A)^a \otimes \beta(B)^b \to A^a \otimes B^b$$

is defined. If necessary, write $\sigma = \sigma_{a,b}$, $\gamma = \gamma_{a,b}$, and $\eta = \eta_{a,b}$. Now, (2.5) implies

(2.9) $$E^p = \bigoplus_{a+b=p} \gamma(A)^a \wedge \beta(B)^b$$

Let

(2.10) $$\pi_{a,b}\colon E^p \to \gamma(A)^a \wedge \beta(B)^b$$

be the projection. Therefore, a linear surjective map

(2.11) $$\rho_{a,b}\colon E^p \to A^a \otimes B^b$$

is defined by $\rho_{a,b} = \eta \circ \sigma \circ \pi_{a,b}$ if $p = a + b$. Observe, that $\rho_{a,b}$ depends on the splitting (2.3). However, if $a = q$, write $\rho = \rho_{q,b}$ for all b. Then ρ is independent of the splitting:

<u>Theorem A II 2.1.</u> <u>If $a = q$ is the fiber dimension of A, the epimorphism</u>

$$\rho = \rho_{q,b}: E^{q+b} \to A^q \otimes B^b$$

<u>does not depend on the splitting (2.3).</u>

<u>Proof</u>. It is sufficient to prove that ρ is independent of the splitting for each single fiber. Hence, it suffices to prove Theorem A II 2.1 for the category of vector space. Hence, suppose that A, B, E are vector spaces with $\dim A = q$, $\dim B = s$ and $\dim E = n$, and that a short exact sequence (2.2) is given. Take a base $(a_1,\ldots,a_q)$ of A and a base $(b_1,\ldots,b_s)$ of B. Suppose that $\gamma(a_1),\ldots,\gamma(a_q)$, $\beta(b_1),\ldots,\beta(b_s)$ are linearly dependent. Then complex numbers c_μ, d_μ exist such that

$$\sum_{\mu=1}^{q} d_\mu \gamma(a_\mu) + \sum_{\mu=1}^{s} d_\mu \beta(b_\mu) = 0.$$

Application of α and ρ implies

$$\sum_{\mu=1}^{q} c_\mu a_\mu = 0 \text{ and } \sum_{\mu=1}^{s} d_\mu b_\mu = 0$$

Hence $c_\mu = 0$ and $d_\mu = 0$ for all indices. Hence (2.5) is proven and $\gamma(a_1),\ldots,\gamma(a_q)$, $\beta(b_1),\ldots,\beta(b_s)$ is a base of E. Take b with $0 \leqq b \leqq s$ and set $p = b + q$. Then

$$\{\gamma(a_\mu) \wedge \beta(b_\nu) \mid \mu \in T(u,q) \text{ and } \nu \in T(v,s)\}$$

is a base of $\gamma(A)^u \wedge \beta(B)^v$ and the union of these sets for $u + v = p$ defines a base of E^p.

Suppose that another splitting

$$(2.12) \qquad 0 \longrightarrow A \xrightarrow{\gamma'} E \xrightarrow{\rho'} B \longrightarrow 0$$

of (2.2) is given. Distinguish the maps associated with the splitting (2.12) by a dash. For $\mu = 1,\ldots,q$

$$\alpha(\gamma(a_\mu) - \gamma'(a_\mu)) = a_\mu - a_\mu = 0 .$$

Hence $e_\mu \in B$ exists such that $\gamma(a_\mu) - \gamma'(a_\mu) = \beta(e_\mu)$. Now, it is claimed that for each $\mu \in T(u,q)$ with $1 \leqq u \leqq q$ an element

$$(2.13) \qquad r_\mu \in \bigoplus_{\substack{j+k=u \\ j<u}} \gamma'(A)^j \otimes \beta(B)^k = C_u$$

exists such that

$$(2.14) \qquad \gamma(a_\mu) = \gamma'(a_\mu) + r_\mu$$

For $u = 1$, this has already been proven. Suppose that it is correct for $u - 1$ with $1 \leqq u - 1 < q$. Take $\mu \in T(u,q)$. Define $\nu = \mu|\Delta_{u-1}$. Then $a_\mu = a_\nu \wedge a_{\mu(u)}$ and $\gamma(a_\nu) = \gamma'(a_\nu) + r_\nu$ with $r_\nu \in G_{u-1}$. Define

$$\begin{aligned} r_\mu &= \gamma(a_\mu) - \gamma'(a_\mu) \\ &= \gamma(a_\nu) \wedge \gamma(a_{\mu(u)}) - \gamma'(a_\nu) \wedge \gamma'(a_{\mu(u)}) \\ &= (-1)^{u-1}\gamma'(a_{\mu(u)}) \wedge r_\nu + r_\nu \wedge \beta(e_{\mu(u)}) \\ &\quad + \gamma'(a_\nu) \wedge \nu(e_{\mu(u)}) \end{aligned}$$

which belongs to C_u.

If $f \in E^p$, then

$$f = \sum_{u+v=p} \sum_{\mu\in T(u,q)} \sum_{\nu\in T(v,s)} f_{\mu\nu}\gamma(a_\mu) \wedge \beta(b_\nu)$$

$$f = \sum_{u+v=p} \sum_{\mu\in T(u,q)} \sum_{\nu\in T(v,s)} f'_{\mu\nu}\gamma'(a_\mu) \wedge \beta(b_\nu).$$

Let ι be the sole element of $T(q,q)$. Then

$$\rho(f) = \sum_{\nu\in T(b,s)} f_{\iota\nu} a_\iota \otimes b_\nu$$

$$\rho'(f) = \sum_{\nu\in T(b,s)} f'_{\iota\nu}\, a_\iota \otimes b_\nu$$

Here $\rho'(r_\mu \wedge \beta(b_\nu)) = 0$ for all $\mu \in T(u,q)$ and $\nu \in T(v,s)$ with $u + v = p$. Therefore

$$\rho'(\gamma'(a_\mu) \wedge \beta(b_\nu)) = \rho'(\gamma(a_\mu) \wedge \beta(b_\nu))$$

if $\mu \in T(u,q)$ and $\nu \in T(v,s)$ with $u + v = p$. Therefore

$$\begin{aligned}\rho'(f) &= \sum_{u+v=p} \sum_{\mu\in T(u,q)} \sum_{\nu\in T(v,s)} f_{\mu\nu}\rho'(\gamma(a_\mu) \wedge \beta(b_\nu)) \\ &= \sum_{u+v=p} \sum_{\mu\in T(u,q)} \sum_{\nu\in T(v,s)} f_{\mu\nu}\rho'(\gamma'(a_\mu) \wedge \beta(b_\nu)) \\ &= \sum_{\nu\in T(b,s)} f_{1\nu}a_1 \otimes b_\nu = \rho(f)\end{aligned}$$

q.e.d.

<u>Lemma A II 2.2.</u> <u>Let M be a manifold of dimension m. Suppose that an exact sequence (2.2) of differentiable vector bundles over M is given with fiber dimension n of E, s of B and q = n - s of A.</u> Let p and t be integers with $q \leqq p \leqq p + t \leqq n$. Take $x \in M$ and $\omega \in E_x^p$ and $\psi \in B_x^t$. <u>Then</u>

$$\rho(\omega \wedge \beta(\psi)) = \rho(\omega) \,\Delta\, \psi$$

<u>Proof</u>. Again it is enough to prove this for category of vector spaces. Take an exact sequence (2.2) and a splitting (2.3). Let $(a_1,\ldots,a_q)$ be a base of A and let $(b_1,\ldots,b_s)$ be a base of B. Then

$$\omega = \sum_{u+v=p} \sum_{\mu \in T(u,q)} \sum_{\nu \in T(v,s)} \omega_{\mu\nu} \gamma(a_\mu) \wedge \beta(b_\nu)$$

$$\rho(\omega) = \sum_{\nu \in T(p-q,p)} \omega_{1\nu} a_1 \otimes b_\nu$$

where 1 is the sole element of $T(q,q)$. Moreover

$$\omega \wedge \beta(\psi) = \sum_{u+v=p} \sum_{\mu \in T(u,q)} \sum_{\nu \in T(v,s)} \omega_{\mu\nu} (a_\mu) \wedge \beta(b_\nu \wedge \psi)$$

Then

$$\rho(\omega \wedge \beta(\psi)) = \sum_{\nu \in T(p-q,q)} \omega_{1\nu} a_1 \otimes (b_\nu \wedge \psi)$$

$$= \rho(\omega) \Delta \psi$$

by (1.20), q.e.d.

§3 Regular maps.

Let M and N be manifolds with dim M = m and dim N = n. Let $f\colon M \to N$ be a differentiable map. Consider a differentiable vector bundle $\pi\colon E \to N$. The pull back is defined by

$$(3.1) \qquad f^*(E) = \{(e,x) \in E \times M \mid f(x) = \pi(e)\}$$

where $\tilde{f}\colon f^*(E) \to E$ and $\tilde{\pi}\colon f^*(E) \to M$ are the natural projections. Here, $\tilde{\pi}\colon f^*(E) \to M$ is a differentiable vector bundle with $f \circ \tilde{\pi} = \pi \circ \tilde{f}$. For each $x \in M$, the map $\bar{f}_x = f \mid f^*(E_x)$ is an isomorphism onto $E_{f(x)}$. The fiber

$$(3.2) \qquad M_a = M_a(f) = f^{-1}(a)$$

of f over a is a closed subset of M. If $M_a \neq \emptyset$, then

$$(3.3) \qquad f^*(E)\mid M_a = \{(e,a) \in E \times M_a \mid a = \pi(e)\} = E_a \times M_a .$$

Hence $f^*(E)\mid M_a = (E_a)_{M_a}$ is the trivial bundle over M_a with general fiber E_a.

Especially, consider the cotangent bundles T(N) of N. It pulls back to $f^*(T(N))$. For every subset U of N, the pull back

$$(3.4) \qquad f^*\colon \Gamma(U,T^p(N)) \to \Gamma(f^{-1}(U),T^p(M))$$

of differentiable forms is defined, and commutes with the restriction maps to subsets of U; for every differential function g on a neighborhood of U, the relation $f^*(dg) = d(g \circ f)$ holds. (f^* is defined by these properties). One and only one vector bundle homomorphism

$$\hat{f}: f^*(T(N)) \to T(M) \tag{3.5}$$

exists such that $(f^*\omega)(x) = \hat{f}(\tilde{f}_x^{-1}(\omega(f(x))))$ for each $x \in U$ and each $\omega \in \Gamma(U,T(N))$. For $x \in M$, define $\hat{f}_x = \hat{f}|f^*(T(N))_x$ and

$$f_x^* = \hat{f}_x \circ \tilde{f}_x^{-1}: T(N)_{f(x)} \to T(M)_x \tag{3.6}$$

The map f is said to be <u>regular</u> (<u>smooth</u>) <u>at $x \in M$</u>, if and only if f_x^* is injective (surjective), which is the case, if and only if $\hat{f}_x$ is injective (surjective). The map f is said to be <u>regular (smooth) on $U \subseteq M$</u>, if and only if f is regular smooth at every $x \in U$. The map f is said to be <u>regular</u> (<u>smooth</u>) if and only if it is so on M. If f is regular (smooth) at $x \in M$, then $m \geqq n$ ($m \leqq n$). The map f is regular (smooth) at $x \in M$, if and only if the Jacobian of f at x has rank n (resp. m). Therefore, the set of regular (smooth) points of f is open in M.

<u>Definition A II 3.1.</u> <u>The couple (α,β) defines a product representation of f if and only if</u>

1.) <u>The maps $\alpha: U_\alpha \to U'_\alpha$ and $\beta: U_\beta \to U'_\beta$ are diffeomorphisms where $U_\alpha \subset M$ and $U'_\alpha \subset \mathbb{R}^m$ and $U_\beta \subset N$ and $U'_\beta \subset \mathbb{R}^n$ are open. (Recall, a diffeomorphism preserves orientation by definition.)</u>

2. <u>Open subsets U''_α of $\mathbb{R}^q$ and U'''_α of U'_β exist such that $U'_\alpha = U''_\alpha \times U'''_\alpha$ where $\pi_\alpha\colon U'_\alpha \to U'''_\alpha$ and $\psi_\alpha\colon U'_\alpha \to U''_\alpha$ are the projections.</u>

3. $\pi_\alpha \circ \alpha = \beta \circ f$.

The product representation is said to be <u>preferred</u> if and only if $U'''_\alpha = U'_\beta$. If $x \in U_\alpha$ (resp. $y \in U'''_\alpha$), the product representation (α,β) is called a product representation at x (resp. over y).

According to the implicit function theorem, a product representation of f at x exists if and only if f is regular at x. Each product representation can be easily changed into a preferred product representation by replacing U'_β by U'''_α and U_β by $\beta^{-1}(U'''_\alpha)$.

Again, let $f\colon M \to N$ be a differentiable map. Suppose that $q = m - n > 0$. Pick $a \in N$ with $f^{-1}(a) \neq \emptyset$. Assume that f is regular on $f^{-1}(a)$. Let $j\colon f^{-1}(a) \to M$ be the inclusion. If (α,β) is a product representation over a, define

$$U_\varepsilon = U_\alpha \cap f^{-1}(a) \quad \text{and} \quad U'_\varepsilon = U''_\alpha$$

(3.7) $$\varepsilon = \varepsilon_{\alpha,\beta} = \varepsilon^a_{\alpha,\beta} = \psi_\alpha \circ \alpha \circ j\colon U_\varepsilon \to U'_\varepsilon$$

Obviously, ε is a topological map.

Let (α',β') be another product representation of f over a with $\varepsilon' = \varepsilon_{\alpha',\beta'}$ and $U_\varepsilon \cap U_{\varepsilon'} \neq \emptyset$. Then

$$(3.8)\quad \begin{aligned} &\alpha = (x_1,\dots,x_m) \qquad \beta = (y_{q+1},\dots,y_m) \\ &\psi_\alpha \circ \alpha = (x_1,\dots,x_q) \qquad \pi_\alpha \circ \alpha = (x_{q+1},\dots,x_m) \\ &x_\mu = y_\mu \circ f \quad \text{for} \quad \mu = q+1,\dots,m \\ &dx_\mu = f^*(dy_\mu) \quad \text{for} \quad \mu = q+1,\dots,m\,. \end{aligned}$$

The same relation with dashes holds for (α',β'). Then

$$(3.9)\qquad \varepsilon' \circ \varepsilon^{-1}(z) = \psi_{\alpha'} \circ \alpha' \circ \alpha^{-1}(z,\beta(a))$$

for $z \in U_\varepsilon \cap U_{\varepsilon'}$. Moreover

$$(3.10)\qquad dy'_\mu = \sum_{\nu=q+1}^{m} Y_{\mu\nu}\, dy_\nu \quad \text{for } \mu = q+1,\dots,m$$

$$(3.11)\qquad dx'_\mu = \sum_{\nu=0}^{m} X_{\mu\nu}\, dx_\nu \quad \text{for } \mu = 1,\dots,m$$

$$(3.12)\qquad dx'_\mu = \sum_{\nu=q+1}^{n} Y_{\mu\nu} \circ f\, dx'_\nu \quad \text{for } \mu = q+1,\dots,m$$

$$(3.13)\qquad X_{\mu\nu} = \begin{cases} 0 & \text{if } 1 \leqq \nu \leqq q \\ Y_{\mu\nu} \circ f & \text{if } q+1 \leqq \nu \leqq m \end{cases}$$

for $\mu = q+1,\dots,m$. The determinants $Y = \det(Y_{\mu\nu})$ and $X = \det(X_{\mu\nu})$ are positive. Define

$$W = \det_{1\leqq\mu,\,\nu\leqq q} X_{\mu\nu}$$

Then $X = W\cdot(Y \circ f)$. Therefore $W > 0$. Observe, that $W(\gamma(z))$ is the

Jacobian of $\varepsilon' \circ \varepsilon^{-1}(z)$. Therefore $\varepsilon' \circ \varepsilon^{-1}$ is a diffeomorphism. Because a product representation of f over a exists at every $x \in f^{-1}(a)$, the set $\mathcal{G}$ of all $\varepsilon_{\alpha,\beta}$ defined by product representations (α,β) of f over a is an atlas for one and only one oriented, differentiable structure of class C^∞ on $f^{-1}(a)$. Hence, $f^{-1}(a)$ becomes a (paracompact, oriented, differentiable) manifold of pure dimension q, and as such, is denoted by $M_a = M_a(f)$. The inclusion map $j: M_a \to M$ is smooth. The construction of M_a has been given in such detail to be sure of the orientation.

Lemma A II 3.2. Let M, N and S be manifolds with dimensions m, n, and s respectively where $m > n > s$. Let $f: M \to N$ and $g: N \to S$ be differentiable and regular maps. Then $h = g \circ f$ is regular. Moreover, pick $c \in S$ with $h^{-1}(c) \neq \emptyset$. Let $j: M_c = M_c(h) \to M$ be the inclusion map. Then $u = f \circ j: M_c \to N_c = N_c(g)$ is regular. If $b \in N_c$, define $(M_c)_b(u) = M_b(u)$. Then $M_b(f) = M_b(u)$.

Proof. For every $x \in M$, the map $h_x^* = f_x^* \circ g_{f(x)}^*$ is injective, because f_x and $g_{f(x)}$ are injective. Hence, h is regular.

Pick $a \in M_c$; preferred product representations (α,β) of f at a and (β,η) of g at $f(a) = b$ exist such that the following commutative diagram holds

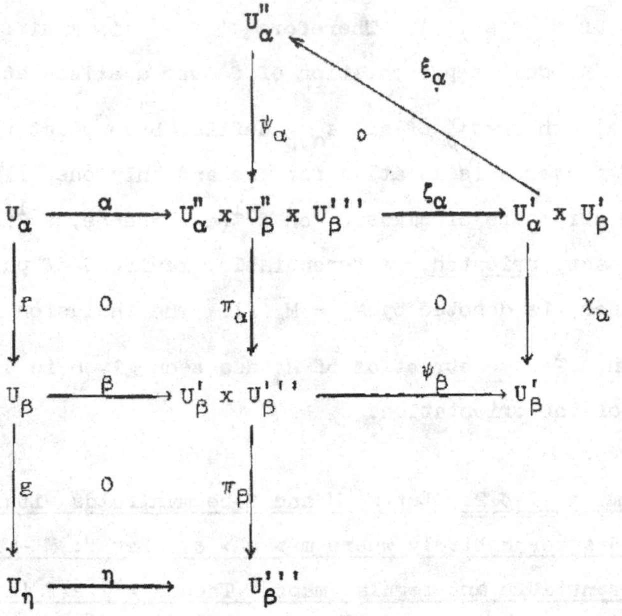

(3.14)

where ζ_α, ξ_α, ψ_α, χ_α, ψ_β, π_α, π_β, and $\tilde{\pi}_\alpha = \pi_\beta \circ \pi_\alpha$ are projections. Let $j\colon M_c \to M$, and $j_b\colon M_b(f) \to M$ and $j_0\colon M_b(u) \to M_c$ be the inclusion map. Because $u^{-1}(b) = f^{-1}(b)$, $j \circ j_0 = j_b$. According to the construction of the structures of $N_c(g)$ and $M_c(h)$, the maps

$$\tilde{\beta} = \psi_\beta \circ \beta \circ j_c\colon U_\beta \cap N_c \to U''_\beta$$

$$\tilde{\alpha} = \zeta_\alpha \circ \alpha \circ j\colon U_\alpha \cap M_c \to U''_\alpha \times U''_\beta$$

are diffeomorphisms with

$$\chi_\alpha \circ \tilde{\alpha} = \chi_\alpha \circ \zeta_\alpha \circ \alpha \circ j = \psi_\beta \circ \beta \circ f \circ j$$
$$= \psi_\beta \circ \beta \circ j_0 \circ u = \tilde{\beta} \circ u$$

Therefore, $(\tilde{\alpha},\tilde{\beta})$ is a product representation of u at a over b. Hence, u is regular. Moreover, $\varepsilon_{\tilde{\alpha},\tilde{\beta}} = \xi_\alpha \circ \tilde{\alpha} \circ j_0$ belongs to the structure of $M_b(u)$, while

$$\varepsilon_{\tilde{\alpha},\tilde{\beta}} = \xi_\alpha \circ \tilde{\alpha} \circ j_0 = \xi_\alpha \circ \zeta_\alpha \circ \alpha \circ j \circ j_0$$

$$= \psi_\alpha \circ \alpha \circ j_b = \varepsilon_{\alpha,\beta}$$

belongs to the structure of $M_b(f)$. Hence $M_b(u) = M_b(f)$, q.e.d.

The concept of a boundary manifold was introduced in §3 page 31. Let f be a C^∞-function on M; then $f: M \to \mathbb{R}$ is regular at $x \in M$ if $(df)(x) \neq 0$. For $r \in \mathbb{R}$, define

$$G_r = \{x \in M \mid f(x) < r\}$$

$$\Gamma_r = \{x \in M \mid f(x) = r\} = f^{-1}(r).$$

<u>Lemma A II 3.3.</u> <u>Let $f: M \to \mathbb{R}$ be a differentiable and regular map. Take $r \in \mathbb{R}$ such that $f^{-1}(r) = \Gamma_r \neq \emptyset$. Then $\Gamma_r = M_r(f)$ is a boundary manifold of G_r if m is odd, and $\Gamma_r = M_r(f)$ is a boundary manifold of $M - G_r$ if m is even.</u>

<u>Proof.</u> Let (α,β) be a product representation at $x_0 \in \Gamma_r$ over r. Then $\alpha = (x_1,\ldots,x_m)$ and $\varepsilon = (x_1,\ldots,x_{m-1})$. The map $\varepsilon: U_\varepsilon \to U'_\varepsilon$ is a diffeomorphism where $U_\varepsilon = U_\alpha \cap \Gamma_r$ and

$U_\alpha \cap G_r = \{z \in U_\alpha | x_n(z) - r < 0\}$. Because $\alpha' = (x_m - r, x_1, \ldots, x_{m-1})$ if m is odd and $\alpha'' = (r - x_m, x_1, \ldots, x_{m-1})$ if m is even, is orientation preserving, the Lemma is proved; q.e.d.

Lemma A II 3.4. Let $f: M \to N$ be a differentiable and regular map. Suppose that $\dim M = m$ and $\dim N = n$ where $q = m - n$ is even. Let H be an open subset of N and let S be a boundary manifold of H. Define $G = f^{-1}(H)$ and $R = f^{-1}(S)$. Then R is a boundary manifold of G. Let $j: R \to M$ be the inclusion. Define $u = f \circ j: R \to S$. Then $R_b(u) = M_b(f)$ for every $b \in S$ with $f^{-1}(b) \neq \emptyset$.

Proof. Because the map f is open, it can be assumed without loss of generality that f is surjective. Pick $b \in S$. Then a diffeomorphism $\beta: U_\beta \to U'_\beta$ of an open neighborhood U_β of b onto an open neighborhood U'_β of $\mathbb{R}^n$ exists such that $\beta = (x_1, \ldots, x_n)$ and

$$U_\beta \cap H = \{z \in U_\beta | x_1(z) < 0\}$$

where $\eta = (x_2, \ldots, x_n): U_\beta \cap S \to U'_\eta$ is a diffeomorphism. Define $W = f^{-1}(U_\beta)$. Define $h = x_1 \circ f$ if n is odd and $h = -x_1 \circ f$ if n is even. In both cases, h is regular with

$$W \cap G = \{z \in W | h(z) < 0\} \quad \text{if m is odd}$$

$$W \cap G = \{z \in W | h(z) > 0\} \quad \text{if m is even.}$$

By Lemma A II 3.3, $W \cap R = W_0(h)$ is a boundary manifold of $W \cap G$ and hence of G. By Lemma A II 3.2, $R_b(u) = M_b(f)$. Pick $b' \in S$ and β' with the same properties as above, such that $U_\beta \cap U_{\beta'} \neq \emptyset$. Because on every relative open subset Y of R at most one oriented, differentiable structure of class C^∞ exists such that Y is a boundary manifold of G, the structures introduced on $W \cap R$ and $W' \cap R$ agree on $W \cap W' \cap R$. Hence, R is a boundary manifold of G, q.e.d.

Let $f: M \to N$ be regular. An exact sequence

$$(3.15) \qquad 0 \longrightarrow f^*(T(N)) \xrightarrow{\hat{f}} T(M) \xrightarrow{r} Q \longrightarrow 0$$

is defined, because $\hat{f}$ is injective. The quotient bundle Q has fiber dimension $q = m - n$ and is called the bundle tangential to the fibers of f. Assume that $q > 0$.

To every product representation of f, a splitting of (3.15) will be assigned: Let (α, β) such a representation and use the notations of Definition A II 3.1 and (3.8). Then $dx_1, \dots, dx_m$ is a frame field of T(M) over U_α and $dy_{q+1}, \dots, dy_m$ is a frame field of T(N) over U_β. A unique frame field $\eta_{q+1}, \dots, \eta_m$ of $f^*(T(N))$ over $f^{-1}(U_\beta) \supseteq U_\alpha$ exists such that $\tilde{f}(\eta_\mu(x)) = (dy_\mu)(f(x))$ for every $x \in f^{-1}(U_\beta)$. Then $dx_\mu = f^*(dy_\mu) = \tilde{f}(\eta_\mu)$ on U_α for $\mu = q + 1, \dots, m$. If $\omega \in T_x(M)$ with $x \in U_\alpha$, then

$$\omega = \sum_{\mu=1}^{m} \omega_\mu dx_\mu .$$

Define $\delta(\omega) = \sum_{\mu=q+1}^{m} \omega_\mu \eta_\mu$. Then $\delta: T(M) \to f^*(T(N))$ is defined over

U_α such that $\delta \circ \hat{f}$ is the identity. Now, $r(dx_1),\ldots,r(dx_q)$ is a frame field of Q over U_α. If $\psi = \sum_{\mu=1}^{q} \psi_\mu r(dx_\mu) \in Q_x$ for $x \in U_\alpha$ define $\gamma(\psi) = \sum_{\mu=1}^{q} \psi_\mu dx_\mu$. Then $r \circ \gamma$ and $\hat{f} \circ \delta + \gamma \circ r$ are identities. Hence a splitting

$$(3.16) \qquad 0 \longrightarrow Q \xrightarrow{\gamma} T(M) \xrightarrow{\delta} f^*(T(N)) \longrightarrow 0$$

over U_α has been associated with (α,β).

<u>Lemma A II 3.5. Let $f\colon M \to N$ be a regular map. Take $a \in N$ with $f^{-1}(a) \neq \emptyset$. Let $j\colon M_a \to M$ be the inclusion. Then</u>

$$(3.17) \qquad 0 \longrightarrow f^*(T(N)) \xrightarrow{\hat{f}} T(M) \xrightarrow{j^*} T(M_a) \longrightarrow 0$$

<u>is exact over M_a.</u>

<u>Proof.</u> Let (α,β) be a product representation of f over a and adopt the previous notations. Define $\varepsilon = \varepsilon_{\alpha,\beta}$. Then $\varepsilon = (x_1 \circ j,\ldots,x_q \circ j)$. Hence, $j^*(dx_1),\ldots,j^*(dx_q)$ is a frame field of $T(M_a)$ over $U_\varepsilon = U_\alpha \cap M_a$. Hence j^* is surjective. Moreover, the kernel of j^* is spanned by $dx_{q+1},\ldots,dx_m$, that is the image of $\hat{f}$, over U_ε. Hence (3.17) is exact over U_ε, q.e.d.

Observe, that Lemma A II 3.5 implies, that $j\colon M_a \to M$ is smooth, which of course could be shown directly. According to Lemma II 3.5, identify

$$(3.18) \qquad Q|M_a = T(M_a) \qquad r|(Q|M_a) = r_a = j^*.$$

Observe that $f^*(T(N))|M_a = T_a(N)_{M_a}$ is the trivial bundle over M_a with general fiber $T_a(N)$. Hence (3.17) and (3.15) becomes

$$(3.19) \qquad 0 \longrightarrow T_a(N)_{M_a} \xrightarrow{\hat{f}} T(M) \xrightarrow{r_a} T(M_a) \longrightarrow 0$$

over M_a.

According to Theorem A II 2.1 a natural homomorphism

$$(3.20) \qquad \rho = \rho^f \colon T(M)^p \to Q^q \otimes f^*(T(N))^s$$

is defined if $s = p - q \geqq 0$. The restriction $\rho_a = \rho_a^f$ to M_a is

$$(3.21) \qquad \rho_a \colon T(M)^p \to T(M_a)^q \otimes T_a(N)_{M_a}$$

if $a \in f(M)$.

Let ω be a form of degree p on M, i.e., $\omega \in \Gamma(M, T(M)^p)$ with $p \geqq q$. Take $a \in N$; if $f^{-1}(a) = \emptyset$, then ω is said to be integrable over the fibers of f at a and $\int_{M_a} \omega = 0$; if $f^{-1}(a) \neq \emptyset$, ω is said to be integrable over the fibers of f at a if and only if $\rho_a(\omega)$ is integrable over M_a, i.e., $\rho_a(\omega) \in D_{M_a}(T_a(N)^s)$. If so, the fiber integral of ω for f at a is defined by

$$(3.22) \qquad (f_*\omega)(a) = \int_{M_a} \rho_a(\omega) \in T_a(N)^s$$

Hence $f_*\omega$ is a form of degree $s = p - q$ on the set of all points a of N where ω is integrable over the fibers of f at a.

Let $j\colon M_a \to M$ be the inclusion. Let (α,β) be a product representation of f over $a \in N$ and at $z \in M_a$. Set $\alpha = (x_1,\ldots,x_m)$ and $\beta = (y_1,\ldots,y_n)$. Then $x_{q+\mu} = y_\mu \circ f$ for $\mu = 1,\ldots,n$. However, $dx_1,\ldots,dx_m$ is a frame field of T(M) over U_α, $dy_1,\ldots,dy_n$ is a frame field of T(N) over U_β and $j^*(dx_1),\ldots,j^*(dx_q)$ is a frame field of $T(M_a)$ over $U_\varepsilon = U_\alpha \cap M_a$. Let ω be a form of degree $p \geqq q$ on M and set $s = p - q$. Then

$$(3.23) \qquad \omega = \sum_{\kappa+\lambda=p} \sum_{\mu\in T(\kappa,q)} \sum_{\nu\in T(\lambda,s)} \omega_{\mu\nu} dx_\mu \wedge d(y_\nu \circ f)$$

$$(3.24) \qquad \rho_a(\omega) = \sum_{\nu\in T(\lambda,s)} (\omega_{1\nu} \circ j) j^*(dx_1) \otimes dy_\nu$$

where 1 is the sole element of T(q,q).

§4 Properties of the fiber integral.

Let M and N be manifolds of dimension m and n respectively with $q = m - n > 0$. Let $f: M \to N$ be a differentiable map. Let $R = R(f)$ be the open set of regular points of f, then $f: R \to N$ is regular, if $R \neq \emptyset$. For $a \in N$, define $M_a = R_a$. Hence $M_a = \emptyset$, if and only if $f^{-1}(a) \cap R = \emptyset$.

The fiber integral was defined at the end of §3 for regular maps. If $f: M \to N$ is only differentiable, a form $\omega \in \Gamma(M,T^p)$ with $q \leqq p \leqq m$ is said to be integrable along the fibers of f at a if and only if ω is integrable along the fibers of $f|R$ at a. If so, define

$$(4.1) \qquad (f_*\omega)(a) = ((f|R)_*\omega)(a)$$

$$= \int_{M_a} \rho_a(\omega) = \int_{R_a} \rho_a(\omega)$$

as the fiber integral.

Lemma A II 4.1. If $\omega_\nu \in \Gamma(M,T^p)$ is integrable over the fibers of f at a for $\nu = 1,\ldots,s$, then $\omega = \omega_1 + \ldots + \omega_s$ is integrable over the fibers of f at a and

$$f_*\omega = f_*\omega_1 + \ldots + f_*\omega_s .$$

Lemma A II 4.2. If $\omega \in \Gamma(M,T^p)$ is integrable over the fibers of f at a, and if $c \in \mathbb{C}$, then $c\omega$ is integrable over the fibers of f at a and

$$f_*(c\omega) = cf_*(\omega).$$

Lemma A II 4.3. If $\omega \in \Gamma(M,T^p)$ and if $\omega|M_a$ is almost everywhere zero on M_a, then ω is integrable over the fibers of f at a and $(f_*\omega)(a) = 0$.

Lemma A II 4.4. Let U be open in N. Take $a \in U$. Define $V = f^{-1}(U)$. Then $\omega \in \Gamma(M,T^p(M))$ is integrable over the fibers of f at a if and only if $\omega|V$ is integrable over the fibers of $f|V: V \to U$ at a and $(f|V)_*(\omega|V) = f_*\omega$ at a.

The proofs of these lemmata follow immediately from the definition, form the fact that ρ_a is a vector bundle homomorphism and from §1.

Lemma A II 4.5. Let $\omega \in \Gamma(M,T^p)$ be integrable over the fibers of f at a. Then $\overline{\omega}$ is integrable over the fibers of f at a and

$$f_*\overline{\omega} = \overline{f_*\omega}$$

Proof. Let (α,β) be a product representation of f at $z \in M_a$. Then $\alpha = (x_1,\ldots,x_m)$ and $\beta = (y_1,\ldots,y_n)$ with $x_{\mu+q} = y_\mu \circ f$ for $\mu = 1,\ldots,n$. Let $j: M_a \to M$ be the inclusion, then

$$(4.2) \qquad \omega = \sum_{\kappa+\lambda=p} \sum_{\mu\in T(\kappa,q)} \sum_{\nu\in T(\lambda,n)} \omega_{\mu\nu}\, dx_\mu \wedge dy_\nu \circ f$$

$$\overline{\omega} = \sum_{\kappa+\lambda=p} \sum_{\mu\in T(\kappa,q)} \sum_{\nu\in T(\lambda,n)} \overline{\omega}_{\mu\nu}\, dx_\mu \wedge dy_\nu \circ f$$

Let 1 be the sole element of $T(q,q)$. Define $s = p - q$. Then

$$(4.3) \qquad \rho_a(\omega) = \sum_{\nu\in T(s,n)} \omega_{1\nu} j^*(dx_1) \otimes dy_\nu$$

$$\rho_a(\overline{\omega}) = \sum_{\nu\in T(s,n)} \overline{\omega}_{1\nu} j^*(dx_1) \otimes dy_\nu = \overline{\rho_a(\omega)}\,.$$

Therefore, $\rho_a(\overline{\omega}) = \overline{\rho_a(\omega)}$. Now, Lemma A II 1.10 completes the proof, q.e.d.

Lemma A II 4.6. Suppose that the form ω of degree $p \geqq q$ on M is integrable over the fibers of f at $a \in N$. Let ψ be a form of degree r on N with $0 \leqq r \leqq m - p$. Then $\omega \wedge f^*\psi$ and $f^*\psi \wedge \omega$ are integrable over the fibers of f at a and

$$f_*(\omega \wedge f^*\psi) = (f_*\omega) \wedge \psi \qquad \text{at } a$$

$$f_*(f^*\psi \wedge \omega) = (-1)^{qr}\psi \wedge f_*\omega \qquad \text{at } a.$$

Proof. According to Lemma A II 2.2

$$\rho_a(\omega \wedge f^*\psi) = \rho_a(\omega) \,\Delta\, \psi(a)$$

According to Lemma A II 1.9 $\rho_a(\omega) \Delta \psi(a)$ is integrable over M_a. Hence $\omega \wedge f^*\psi$ is integrable over the fibers of f at a and

$$f_*(\omega \wedge f^*\psi) = \int_{M_a} \rho_a(\omega) \Delta \psi(a) = (\int_{M_a} \rho_a(\omega)) \wedge \psi(a)$$

$$= f_*(\omega) \wedge \psi \qquad \text{at a.}$$

Now

$$f_*(f^*\psi \wedge \omega) = (-1)^{pr} f_*(\omega \wedge f^*\psi) = (-1)^{pr} f_*\omega \wedge \psi$$

$$= (-1)^{pr+(p-q)r} \psi \wedge f_*\omega = (-1)^{qr} \psi \wedge f_*\omega.$$

Let L be a smooth submanifold of M and let $j: L \to M$ be the inclusion map. Suppose that L has dimension q. Let ω be a form of degree q on M. Suppose that $j^*(\omega)$ is integrable over L. Then $\int_L \omega = \int_L j^*(\omega)$ by definition. Observe that M_a is a smooth, q-dimensional submanifold of M. Hence this situation applies.

<u>Lemma A II 4.7.</u> <u>Let ω be a form of degree q on M. Then ω is integrable over the fibers of f at a if and only if ω (i.e., $j^*(\omega)$) is integrable over M_a. If so, then</u>

$$(f_*\omega)(a) = \int_{M_a} \omega .$$

<u>Proof.</u> As (3.23) and (3.24) show $j^*(\omega) = \rho_a(\omega)$ in this case, q.e.d.

Lemma A II 4.8. Let ω be a form of degree $p \geqq q$ on M. Assume that $\omega|M_a$ is continuous and has compact support on M_a, then ω is integrable over the fibers of f at a.

Proof. Obviously, $\rho_a(\omega)$ is continuous and its support is contained in the support of $\omega|M_a$. Therefore $\rho_a(\omega)$ has compact support on M_a. Hence $\int_{M_a} \rho_a(\omega)$ exists; q.e.d.

Lemma A II 4.9. Let ω be a form of class C^k and degree $p \geqq q$ on M with $0 \leqq k \leqq \infty$. Let K be the support of ω. Suppose that $f|K: K \to N$ is proper. Suppose that f is regular on K. Then $f_*\omega$ exists for every point of N and is a form of class C^k on N.

Proof. Without loss of generality, it can be assumed that $f: M \to N$ is regular. Because $M_a \cap K$ is compact for every $a \in N$, the fiber integral $f_*\omega$ exists for every $a \in M$ by Lemma A II 4.8. Then $f_*\omega$ is a form of degree $s = p - q$ on N.

1. Special case. Suppose that $M = S \times N$, where N is open in $\mathbb{R}^n$ and S is open in $\mathbb{R}^q$. Suppose that $f: M \to N$ is the projection. Let $\psi: M \to S$ be the projection. Let $x_1, \ldots, x_q$ be the coordinates on S and $y_1, \ldots, y_n$ the coordinates on N. Then $x_1 \circ \psi, \ldots, x_q \circ \psi$, $y_1 \circ f, \ldots, y_n \circ f$ are the coordinates of M. Set $s = p - q$. On M

$$\omega = \sum_{\kappa+\lambda=p} \sum_{\mu\in T(\kappa,q)} \sum_{\nu\in T(\lambda,n)} \omega_{\mu\nu}\psi^*(dx_\mu) \wedge f^*(dy_\nu)$$

where $\omega_{\mu\nu}$ are functions of class C^k on M. If $y \in N$, and if ι is the sole element of $T(q,q)$, then

$$\rho_y(\omega)(x) = \sum_{\nu \in T(s,n)} \omega_{\iota\nu}(x,y)\, dx_\iota \otimes dy_\nu .$$

Take $a \in N$ and let U be an open, relative compact neighborhood of N. Then $K_1 = K \cap f^{-1}(\overline{U})$ is compact. The fiber integral is given by

$$(f_*\omega)(y) = \sum_{\nu \in T(s,n)} \left(\int_{K_1} \omega_{\iota\nu}(x,y)\, dx_\iota\right) dy_\nu$$

for $y \in U$. Hence $f_*\omega$ is of class C^k on U.

The special case implies that the Lemma is true, always if K is compact and contained in U_α for some product representation (α,β) of f, because $f_*(\omega)$ has compact support in U_β and is of class C^k in U_β.

2. The general case. Take $a \in N$. Let U be an open, relative compact neighborhood of a. Then $K_1 = K \cap f^{-1}(\overline{U})$ is compact. Finitely many product representations (α_ν, β_ν) $\nu = 1,\ldots,t$ exist such that $K_1 \subseteq \bigcup_{\nu=1}^{t} U_{\alpha_\nu}$. A partition of unity $\{g_\nu\}$ $\nu = 1,\ldots,t$ by C^∞-functions on M exists such that g_ν has compact support in U_{α_ν} and such that $\sum_{\nu=1}^{t} g_\nu(x) = 1$ if $x \in K_1$. Then $f_*(g_\nu\omega)$ is of class C^k on N. Define $g = g_1 + \ldots + g_t$. Then

$$f_*(g\omega) = \sum_{\nu=1}^{t} f_*(g_\nu\omega)$$

has class C^k on N. Because $g\omega = \omega$ on $f^{-1}(U)$, this implies that $f_*\omega|U = f_*(g\omega)|U$ has class C^k on U; q.e.d.

Lemma A II 4.10.[42] Let M and N manifolds of dimension m and n respectively with $q = n - n > 0$. Let $f\colon M \to N$ be a differentiable and regular map. Let ω be a form of degree m on M. Let $\beta\colon U_\beta \to U'_\beta$ be a diffeomorphism of the open subset U_β of N onto the open subset U'_β of $\mathbb{R}^n$. Define $W_\beta = f^{-1}(U_\beta)$. Set $\beta = (y_1,\ldots,y_n)$. Then, for every $a \in U_\beta$, one and only one form $\omega_a^\beta \in \Gamma(M_a, T^q(M_a))$ of degree q on M_a exists such that

$$\rho_a(\omega) = \omega_a^\beta \otimes dy_1 \wedge \ldots \wedge dy_n. \tag{4.4}$$

Moreover, the following properties hold:

a) ω is integrable over the fibers of f at $a \in U_\beta$ if and only if ω_a^β is integrable over M_a and

$$(f_*\omega)(a) = \left(\int_{M_a} \omega_a^\beta\right) dy_1 \wedge \ldots \wedge dy_n \tag{4.5}$$

b) If $\omega \geqq 0$, then $\omega_a^\beta \geqq 0$ on M_a and $(f_*\omega)(a) \geqq 0$ provided this fiber integral exists.

c) Let $a \in U_\beta$. Suppose that ω_a^β is measurable on M_a. Then $|\omega|$ is integrable over the fibers of f at a if and only if ω is integrable over the fibers of f at a. If so, then

$$|f_*\omega| \leqq f_*|\omega| \tag{4.6}$$

where

$$\rho_a(|\omega|) = |\omega_a^\beta| \otimes dy_1 \wedge \dots \wedge dy_n . \tag{4.7}$$

Proof. Since $dy_1 \wedge \dots \wedge dy_n$ is a frame of $T^n(N)$ over U_β, and because $\rho_a(\omega)$ is a section over $M_a \subseteq W_\beta$ of $T^q(M_a) \otimes T_a^n(N)_{M_a}$, (4.4) holds with unique $\omega_a^\beta \in \Gamma(M_a, T^q(M_a))$, which implies a) immediately.

Let $j_a \colon M_a \to M$ be the inclusion. At every $z_0 \in M_a$ a product representation (α,β) of f exists. Define $\alpha = (x_1,\dots,x_m)$. Then $dx_{q+\mu} = f^*(dy_\mu)$ for $\mu = 1,\dots,n$. On U_α

$$\omega = g_{\alpha\beta}\, dx_1 \wedge \dots \wedge dx_m.$$

where $g_{\alpha\beta} \geqq 0$ if and only if $\omega \geqq 0$. On $U_\alpha \cap M_a$:

$$\omega_a^\beta = g_{\alpha\beta} j_a^*(dx_1 \wedge \dots \wedge dx_q)$$

Hence $\omega \geqq 0$, implies $\omega_a^\beta \geqq 0$ and $f_*(\omega)(a) \geqq 0$ if this fiber integral exists. Moreover,

$$|\omega| = |g_{\alpha\beta}|\, dx_1 \wedge \dots \wedge dx_m.$$

Hence (4.7) holds. Hence if ω_a^β is measurable on M_a, then ω_a^β is integrable over M_a if and only if $|\omega_a^\beta|$ is integrable over M_a, meaning ω is integrable over the fibers of f at a, if and only if $|\omega|$ is integrable over the fibers of f at a. If so, then

$$\begin{aligned} |(f_*\omega)(a)| &= |\int_{M_a} \omega_a^\beta|\, dy_1 \wedge \dots \wedge dy_n \\ &\leq \int_{M_a} |\omega_a^\beta|\, dy_1 \wedge \dots \wedge dy_n \\ &= (f_*|\omega|)(a) \qquad \text{q.e.d.} \end{aligned}$$

<u>Theorem A II 4.11.</u>[42] <u>Let M and N be manifolds of dimensions m and n respectively. Assume that $q = m - n > 0$. Let $f: M \to N$ be a differentiable map. Suppose that f is regular at almost every point of M. Let $\omega \in D_M$ be a form of degree m which is integrable over M. Then ω is integrable over the fibers of f at almost every $a \in N$ and $f_*\omega$ is integrable over N with</u>

$$(4.8) \qquad \int_M \omega = \int_N f_*\omega$$

<u>Proof.</u> 1. <u>Special Case</u>: Suppose that $M = S \times N$, where N is open in $\mathbb{R}^n$ and S is open in $\mathbb{R}^q$. Suppose that $f: M \to N$ is the projection. Let $\psi: M \to S$ be the projection. Let $x_1, \dots, x_q$ be the coordinates on S and $y_1, \dots, y_n$ the coordinates on N. Then $x_1 \circ \psi, \dots, x_q \circ \psi, \dots, y_n \circ f$ are the coordinates on M. Then

$$(4.9) \qquad \omega = g\psi^*(dx_1 \wedge \dots \wedge dx_q) \wedge f^*(dy_1 \wedge \dots \wedge dy_n)$$

where g is integrable over M. Hence

$$\int_M \omega = \int_{S\times N} g(x,y)\, dx_1 \wedge \dots \wedge dx_q \wedge dy_1 \wedge \dots \wedge dy_n$$

(4.10)

$$= \int_N \left(\int_S g(x,y)\, dx_1 \wedge \dots \wedge dx_q\right) dy_1 \wedge \dots \wedge dy_n$$

where the interior integral

(4.11) $$(f_*\omega)(y) = \int_S g(x,y)\, dx_1 \wedge \dots \wedge dx_q$$

exists for almost all $y \in N$ and is integrable over M.

If (α,β) is a product representation of the regular map f and if ω has compact support in U_α, the theorem is true according to special case 1.

2. <u>Special case</u>: Let $f: M \to N$ be regular. Suppose that ω has compact support K. Then finitely many product representations (α_ν, β_ν) $\nu = 1,\dots,t$ exist such that

$$K \subseteq \bigcup_{\nu=1}^{t} U_{\alpha_\nu}.$$

Take a partition of unity of C^∞ functions g_ν such that $g = g_1 + \dots + g_t$ equals 1 on K and such that g_ν has compact support in U_{α_ν}. Then $g\omega = \omega$ on M. Then $f_*(g_\nu\omega)$ exists for almost all $a \in N$ and is integrable over N. The same is true for the sum

$$f_*(\omega) = \sum_{\nu=1}^{t} f_*(g_\nu\omega)$$

and

$$\int_M \omega = \sum_{\nu=1}^{t} \int_M g_\nu \omega = \sum_{\nu=1}^{t} \int_N f_*(g_\nu \omega) = \int_N f_* \omega.$$

Therefore, the theorem is true in this special case.

3. <u>Special case</u>: Let $\beta: U_\beta \to U'_\beta$ be a diffeomorphism of an open subset U_β on N onto an open subset U'_β of R^n. Define $(y_1,\ldots,y_n) = \beta$. Suppose that $f: M \to N$ is regular. Suppose that f (supp ω) is contained in a compact subset of U_β.

Let $\{K_\lambda\}_{\lambda \in N}$ be a sequence of compact subset of $W_\beta = f^{-1}(U_\beta)$ such that $K_\lambda \subseteq K_{\lambda+1}$ for all $\lambda \in \mathbb{N}$ and such that $W_\beta = \bigcup_{\lambda=1}^{\infty} K_\lambda$. Let χ_λ be the characteristic function of K_λ. Then $0 \leqq \chi_\lambda \leqq \chi_{\lambda+1} \leqq 1$ and $\chi_\lambda \to 1$ for $\lambda \to \infty$ on W_β.

For $a \in U_\beta$, the representation (4.4) holds. Moreover,

$$\infty > \int_M |\omega| = \lim_{\lambda \to \infty} \int_{W_\beta} \chi_\lambda |\omega| = \lim_{\lambda \to \infty} \int_{U_\beta} f_*(\chi_\lambda |\omega|)$$

where

$$f_*(\chi_\lambda |\omega|)(a) = \int_{M_a} \chi_\lambda |\omega_a^\beta| \, dy_1 \wedge \ldots \wedge dy_n$$

Now,

$$\int_{M_a} \chi_\lambda |\omega_a^\beta| \to \int_{M_a} |\omega_a^\beta| \leqq \infty \qquad \text{for } \lambda \to \infty.$$

Hence

$$\infty > \int_M |\omega| = \int_{a \in U_\beta} (\int_{M_a} |\omega_a^\beta|)\, dy_1 \wedge \dots \wedge dy_n$$
$$= \int_{U_\beta} f_*(|\omega|) = \int_N f_*(|\omega|).$$

Hence $f_*(|\omega|)$ exists almost everywhere on U_β (hence on N) and is integrable over U_β (hence over N). Because of special case 1, ω_a^β is measurable on M_a for almost all $a \in U_\beta$. Therefore $f_*(\omega)$ exists for almost all $a \in U_\beta$ and all $a \in N - U_\beta$ and $|f_*(\chi_\lambda \omega)| \leqq f_*(|\omega|)$ by (4.6). Hence

$$\int_M \omega = \lim_{\lambda \to \infty} \int_{W_\beta} \chi_\lambda \omega = \lim_{\lambda \to \infty} \int_{U_\beta} f_*(\chi_\lambda \omega)$$
$$= \int_{U_\beta} \lim_{\lambda \to \infty} f_*(\chi_\lambda \omega) = \int_{U_\beta} f_* \omega$$
$$= \int_N f_*(\omega)$$

because

$$f_*(\chi_\lambda \omega) = (\int_{M_a} \chi_\lambda \omega_a^\beta)\, dy_1 \wedge \dots \wedge dy_n$$
$$\to \int_{M_a} \omega_a^\beta\, dy_1 \wedge \dots \wedge dy_n = f_*(\omega)$$

for almost all $a \in U_\beta$. The theorem is proved in this case.

4. Special case: Suppose that $f: M \to N$ is regular. Assume that f (supp ω) is contained in a compact subset K of N.

Finitely many diffeomorphisms $\beta_\nu: U_{\beta_\nu} \to U'_{\beta_\nu}$ of an open subset U_{β_ν} of N onto an open subset U'_{β_ν} of $\mathbb{R}^n$ exist such that $K \subseteq U_{\beta_1} \cup \ldots \cup U_{\beta_t}$. Take a partition of unity by C^∞-functions g_ν on N such that g_ν has compact support which is contained in U_{β_ν} and such that $g = g_1 + \ldots + g_t$ with $g = 1$ on K. Then $f_*(g_\nu \circ f\omega) = g_\nu f_*(\omega)$ exist almost everywhere on N and is integrable over N. Then

$$f_*(\omega) = f_*((g \circ f)\omega) = \sum_{\nu=1}^{t} f_*((g_\nu \circ f)\omega)$$

$$= \sum_{\nu=1}^{t} g_\nu f_*(\omega) = gf_*(\omega)$$

exists almost everywhere on N and is integrable over N, with

$$\int_M \omega = \sum_{\nu=1}^{t} \int_M g_\nu \circ f\omega = \sum_{\nu=1}^{n} \int_N f_*(g_\nu \circ f\omega)$$

$$= \sum_{\nu=1}^{t} \int_N g_\nu f_*\omega = \int_N f_*\omega.$$

The Theorem is proven in this case.

5. <u>Special case:</u> $f: M \to N$ is regular. Then take a sequence of compact sets K_λ with $K_\lambda \subseteq K_{\lambda+1} \subseteq N$ and $N = \bigcup_{\lambda=1}^{\infty} K_\lambda$. Let χ_λ be the characteristic function of K_λ. Then $0 \leqq \chi_\lambda \leqq \chi_{\lambda+1} \leqq 1$ and $\chi_\lambda \to 1$ for $\lambda \to \infty$. Now,

$$\infty > \int_M |\omega| = \lim_{\lambda \to \infty} \int_M (\chi_\lambda \circ f)|\omega|$$

$$= \lim_{\lambda \to \infty} \int_N \chi_\lambda f_*(|\omega|)$$

$$= \lim_{\lambda \to \infty} \int_{K_\lambda} f_*(|\omega|) = \int_N f_*(|\omega|)$$

because $f_*(|\omega|) \geqq 0$. Here $f_*(|\omega|)$ exists almost everywhere and is integrable over N. Hence $f_*(\omega)$ exists almost everywhere and is integrable over N. Hence

$$\int_M \omega = \lim_{\lambda \to \infty} \int_M (\chi_\lambda \circ f)\omega = \lim_{\lambda \to \infty} \int_N \chi_\lambda f_*(\omega)$$

$$= \lim_{\lambda \to \infty} \int_{K_\lambda} f_*(\omega) = \int_N f_*(\omega).$$

The Theorem is proved in this case.

6. <u>The general case:</u> Let R be the set of regular point of f in M. Then R is open and M - R has measure zero. Moreover, $(f|R)_*\omega = f_*\omega$. Hence $f_*\omega$ exists almost everywhere on N and is integrable over N. Therefore,

$$\int_M \omega = \int_R \omega = \int_N f_*\omega$$

q.e.d.

Theorem A II 4.11 is the base for the distribution treatment of the integration over the fibers. Observe, that Theorem A II 4.11 is a version of the Fubini theorem. Hence, its inverse also holds:

Theorem A II 4.12.[43)] Let M and N be manifolds of dimensions m and n respectively. Assume that $q = m - n \geqq 0$. Let $f: M \to N$ be a differentiable map. Suppose that f is regular at almost every point of M. Let ω be a non-negative measurable form of degree m on M. Suppose that ω is integrable over almost all fibers of f and that $f_*\omega$ is integrable over N. Then ω is integrable over M and (4.8) holds.

Proof. At first make the same assumptions about $M = S \times N$, S, N and f as in the proof of Theorem A II 4.11, 1. special case. Then (4.9) holds with a non-negative, measurable function g. By assumption, (4.11) exists for almost all $y \in N$ and is integrable over N. Moreover, $g \geqq 0$. Now (4.10) shows that ω is integrable over M.

In the general case, this implies that ω is locally integrable on the set of regular points of M, hence on M. Therefore, ω is integrable over any compact subset K of ω. Let $\{K_\lambda\}_{\lambda \in \mathbb{N}}$ be an increasing sequence of compact sets on M, such that $M = \bigcup_{\lambda=1}^{\infty} K_\lambda$. Let χ_λ be the characteristic function of K_λ. Then

$$0 \leqq \int_M \omega = \lim_{\lambda \to \infty} \int_{K_\lambda} \omega = \lim_{\lambda \to \infty} \int_M \chi_\lambda \omega$$

$$= \lim_{\lambda \to \infty} \int_N f_*(\chi_\lambda \omega) \leqq \int_N f_*\omega < \infty$$

by Lemma A II 4.10 b. Therefore, ω is integrable over M, q.e.d.

Let $f: M \to N$ and $g: N \to S$ be proper, differentiable and regular maps. Let $q > 0$ be the fiber dimension of f and let $r > 0$ be the fiber dimension of g. Let ω be a continuous form of degree $p \geqq r + q$ on M. Then $f_*\omega$ is a continuous form of degree $p - q \geqq r$ on N. Hence $g_*f_*\omega$ is a continuous form of degree $p - q - r$ on N. Now, $g_*f_*\omega = (g \circ f)_*\omega$ is claimed.

For, take a C^∞-form φ degree $m - p = n - (p-q-r)$ on S with compact support. Set $h = g \circ f$. Then

$$\int_S h_*\omega \wedge \varphi = \int_S h_*(\omega \wedge h^*\varphi) = \int_M \omega \wedge h^*\varphi$$

$$= \int_M \omega \wedge f^*g^*\varphi = \int_N f_*\omega \wedge g^*\varphi$$

$$= \int_N g_*f_*\omega \wedge \varphi$$

Because this is true for any such φ, the claim $h_*\omega = g_*f_*\omega$ follows. This result shall be proven in more generality by other methods. For this, it is convenient to extend the concept of a fiber integral:

Let V be a complex vector space of dimension t. Let $f: M \to N$ be a differentiable map of the m-dimensional manifold M into the n-dimensional manifold N with $q = m - n > 0$. Take a form

$\omega \in \Gamma(M, T^p(M) \otimes V_M)$ with $p \geqq q$. Let $e = (e_1, \ldots, e_t)$ be a base of V over $\mathbb{C}$. Then

$$(4.12) \qquad \omega = \sum_{\nu=1}^{t} \omega_\nu \otimes e_\nu$$

with $\omega_\nu \in \Gamma(M, T^p(M))$. Then ω is said to be <u>integrable over the fibers of f at a $\in$ N</u> if and only if each ω_μ is integrable over the fibers of f at a. If so, define

$$(4.13) \qquad (f_*\omega)(a) = \sum_{\nu=1}^{t} (f_*\omega_\nu)(a) \otimes e_\nu \in T_a^{p-q}(N) \otimes V.$$

Obviously, this definition does not depend on the choice of the base e. Also this definition shows that Lemma A II 4.1 - Lemma A II 4.4, Lemma A II 4.7 - Lemma A II 4.9 and Theorem A II 4.12 remain true. If V is a vector space with conjugation, also Lemma A II 4.5 holds. For instance for Theorem A II 4.12 this reads:

<u>Theorem A II 4.13.</u> <u>Let M and N be manifolds of dimensions m and n respectively. Assume that $q = m - n > 0$. Let $f: M \to N$ be a differentiable map. Suppose that f is regular at almost every point of M. Let V be a complex vector space. Let $\omega \in D_M(V)$ be a form of degree m which is integrable over M. Then ω is integrable over the fibers of f at almost every a $\in$ N and $f_*\omega$ is integrable over N with</u>

$$\int_M \omega = \int_N f_*\omega.$$

Proof. Let $e = (e_1,\ldots,e_t)$ be a base of V. Then (4.12) holds, where $\omega_\nu \in D_M$ for $\nu = 1,\ldots,t$. Hence $f_*(\omega_\nu)$ exists almost everywhere on N and is integrable over N. By (4.13), also $f_*\omega$ exists almost everywhere on N and is integrable over N with

$$\int_M \omega = \sum_{\nu=1}^{t} (\int_M \omega_\nu) e_\nu = \sum_{\nu=1}^{n} (\int_N f_*\omega_\nu) e_\nu$$

$$= \int_N \sum_{\nu=1}^{n} f_*\omega_\nu \otimes e_\nu = \int_N f_*\omega$$

q.e.d.

Lemma A II 4.14. Let M, N and S be manifolds with dimensions m,n and s respectively. Suppose that $q = m - n > 0$ and $r = n = s > 0$. Let $f: M \to N$ and $g: N \to S$ be differentiable maps. Define $h = g \circ f$. Take $c \in S$ with $M_c = M_c(h) \neq \emptyset$. Let $j: M_c \to M$ be the inclusion. Then $u = f \circ j: M_c \to N_c$ is a regular map. Take $b \in N_c$. Let ω be a form of degree $p \geqq q + r$ on M, which is integrable over the fibers of f at b. Then

$$\rho_c^h(\omega) \in \Gamma(M_c, T^{q-+\nu}(M_c) \otimes T_c{}^{p-q-\nu}(S)_{M_c})$$

is integrable over the fibers of u at b and

$$\rho_c^g f_*(\omega) = u_* \rho_c^h(\omega) \qquad \text{at } b. \tag{4.14}$$

Proof. Let R_f, R_g and R_h be the set of regular points of f, g and h respectively. For $x \in M$,

$$h_*^* = f_x^* \circ g_{f(x)}^* : T_{h(x)}(S) \to T_{f(x)}(N) \to T_x(M)$$

is injective, if and only if f_x^* and $g_{f(x)}^*$ are injective. Hence

$$R_h = R_f \cap f^{-1}(R_g).$$

Therefore,

$$M_c = R_h \cap h^{-1}(c) = R_h \cap f^{-1}(g^{-1}(c))$$

$$= R_f \cap f^{-1}(g^{-1}(c) \cap R_g) = R_f \cap f^{-1}(N_c) .$$

By Lemma A II 3.2, the map $u = f \circ j\colon M_c \to N_c$ is regular. Abbreviate $M_b(u) = (M_c(h))_b(u)$. Then $M_b(u) = M_b(f)$. Define $p - q = d$ and $p - q - r = e$.

Let (β, η) be a product representation of g at b. Then $\beta = (y_1, \ldots, y_n)$ and $\eta = (z_1, \ldots, z_s)$ with $y_{r+\mu} = z_\mu \circ g$ for $\mu = 1, \ldots, s$. Then

$$\{dy_\mu \wedge g^* dz_\nu \mid (\mu, \nu) \in T(\varphi, r) \times T(\psi, s), \varphi + \psi = d\}$$

is a frame field of $T^d(N)$ over U_β. Therefore

$$\rho_b^f(\omega) = \sum_{\varphi+\psi=d} \sum_{\mu \in T(\varphi, r)} \sum_{\nu \in T(\psi, s)} \omega_{\mu\nu} \otimes dy_\mu \wedge g^* dz_\nu$$

where each $\omega_{\mu\nu}$ is integrable over $M_f(b) = M_u(b)$. Moreover,

$$f_*(\omega)(b) = \sum_{\varphi+\psi=d} \sum_{\mu\in T(\varphi,r)} \sum_{\nu\in T(\psi,s)} \left(\int_{M_b(f)} \omega_{\mu\nu}\right) dy_\mu \wedge g^*(dz_\nu).$$

Now, let κ be the only element of $T(r,r)$, then

$$\rho_c^g f_*(\omega)(b) = \sum_{\nu\in T(d,s)} \left(\int_{M_f(b)} \omega_{\kappa\nu}\, dy_\kappa\right) \otimes dz_\nu.$$

Take $a \in M_f(b)$. Let (α,β) be a product representation of f at a. Then $\alpha = (x_1,\ldots,x_m)$ with $x_{q+\mu} = y_\mu \circ f$ for $\mu = 1,\ldots,n$. On u_α

$$\omega = \sum_{\varphi+\psi+\chi=p} \sum_{\lambda\in T(\chi,q)} \sum_{\mu\in T(\varphi,r)} \sum_{\nu\in T(\psi,r)} \omega_{\lambda\mu\nu}\, dx_\lambda \wedge f^*(dy_\mu) \wedge f^*(dz_\nu)$$

Let ι be the only element of $T(q,q)$. Then

$$\rho_b^f(\omega) = \sum_{\varphi+\psi=d} \sum_{\mu\in T(\varphi,r)} \sum_{\nu\in T(\psi,s)} \omega_{\iota\mu\nu}\, dx_\iota \otimes (dy_\mu \wedge g^*(dz_\nu))$$

on $U_\alpha \cap M_f(b)$. Hence

$$\omega_{\mu\nu} = \omega_{\iota\mu\nu}\, dx_\iota \qquad \text{on } U_\alpha \cap M_f(b).$$

Moreover,

$$\rho_c^h(\omega) = \sum_{\nu\in T(e,s)} \omega_{\iota\kappa\nu}\, dx_\iota \wedge u^*(dy_\kappa) \otimes dz_\nu$$

on $U_\alpha \cap M_c$. Hence

$$\rho_c^h(\omega) = \sum_{\nu\in T(e,s)} \omega_{\kappa\nu}\, u^*(dy_\kappa) \otimes dz_\nu$$

on M_c, where $\omega_{\kappa\nu}$ is integrable over $M_b(f) = M_b(u)$. Hence, $\rho_c^h(\omega)$ is integrable over the fibers of u at b and

$$u_*\rho_c^h(\omega)(b) = \sum_{\nu\in T(e,s)} \left(\int_{M_b(u)} \omega_{\kappa\nu}\, dy_\kappa\right) \otimes dz_\nu$$

$$= \rho_c^g f_*(\omega)(b)$$

q.e.d.

Theorem A II 4.15. Let M, N and S be manifolds with dimensions m, n and s respectively. Suppose that $q = m - n > 0$ and $r = n - s > 0$. Let $f: M \to N$ and $g: N \to S$ be differentiable maps. Define $h = g \circ f$. Take $c \in S$. Let ω be a form of degree $p \geq q + r$ on M, which is integrable over the fibers of f at almost every point of $N_c(g)$. Let ω be integrable over the fibers of h at c. Then $f_*\omega$ is integrable over the fibers of g at c and

$$h_*\omega = g_* f_*\omega \qquad \text{at } c.$$

Proof. The notations of the proof of Lemma A II 4.14 are used. The theorem is trivial if $M_c(h) = \emptyset$. Assume $M_c(h) \neq \emptyset$. Then (4.14) holds at almost every $b \in N_c(g)$. Hence

$$(h_*\omega)(c) = \int_{M_c(h)} \rho_c^h(\omega) = \int_{N_c(g)} u^*(\rho_c^h(\omega))$$

$$= \int_{N_c(g)} \rho^g f_*(\omega) = g^* f_*(\omega)(c)$$

and $f_*(\omega)$ is integrable over the fibers of g at c, q.e.d.

Theorem A II 4.16. Let M, N, P, S be manifolds of dimensions m, n, p and s respectively, such that $q = m - n = p - s > 0$. Let f, g, u and v differentiable maps such that the following diagram commutes:

$$(4.15)\qquad \begin{array}{ccc} P & \xrightarrow{\ u\ } & M \\ \Big\downarrow g & \circ & \Big\downarrow f \\ S & \xrightarrow{\ v\ } & N \end{array}$$

Suppose that g and f are regular. Take $a \in S$. Define $b = v(a)$. Let $j: M_b(f) \to M$ and $k: P_a(g) \to P$ be the inclusion maps. Then one and only one map $u_a: P_a(g) \to M_b(f)$ exists such that $u \circ k = j \circ u_a$. Suppose that u_a is a diffeomorphism.

Let ω be a form of degree $t \geqq q$ on M which is integrable over the fibers of f at b. Then $u^*(\omega)$ is integrable over the fibers of g at a. Moreover,

$$(4.16)\qquad g_* u^* \omega = v^* f_* \omega \qquad \text{at } a.$$

Proof. Take $c \in P_a(g)$. Define $d = u(c)$. Define $r = q - t$. Then

$$(4.17)\qquad \begin{array}{ccc} T_d^p(M) & \xrightarrow{\ \tilde{\rho} = \rho_b^f\ } & T_d^q(M_b(f)) \otimes T_b^r(N)_{M_b(f)} \\ \Big\downarrow u^* & & \Big\downarrow u_a^* \otimes v^* \\ T_c^p(P) & \xrightarrow{\ \rho = \rho_a^g\ } & T_c^q(P_a(g)) \otimes T_a^r(S)_{P_a(g)} \end{array}$$

Now, it shall be proved that (4.17) is commutative.

Let (α,β) be a product representation of f at d. Let (ξ,ζ) be a product representation of g at c such that $v(U_\zeta) \subseteq U_\beta$ and $u(U_\xi) \subseteq U_\alpha$. The following diagram is commutative:

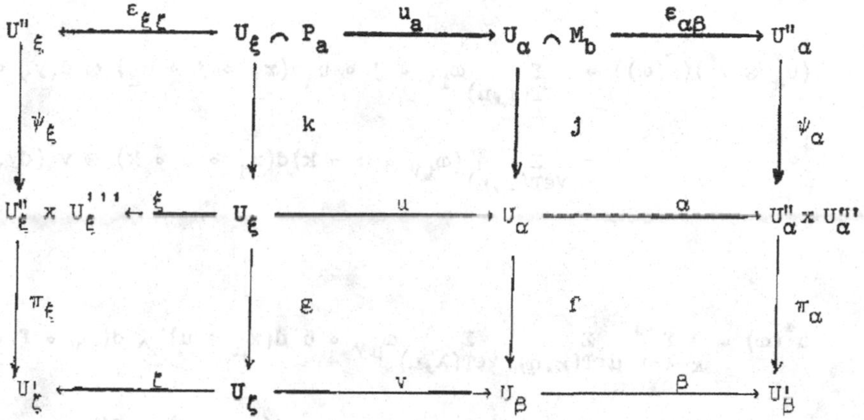

where ψ_ξ, ψ_α, π_ξ and π_α are the projections. Moreover,

$$\alpha = (x_1,\ldots,x_m) \qquad \beta = (y_1,\ldots,y_n)$$

$$\xi = (w_1,\ldots,w_p) \qquad \zeta = (z_1,\ldots,z_s)$$

where $x_{q+\mu} = y_\mu \circ f$ for $\mu = 1,\ldots,n$ and $w_{q+\mu} = z_\mu \circ g$ for $\mu = 1,\ldots,s$. Let i be the sole element of $T(q,q)$. Then

$$\omega = \sum_{\kappa+\lambda=t} \sum_{\mu\in T(\kappa,q)} \sum_{\nu\in T(\lambda,n)} \omega_{\mu\nu}\, dx_\mu \circ f^*(dy_\nu)$$

$$\tilde{\rho}(\omega) = \sum_{\nu\in T(r,n)} (\omega_{1\nu} \circ j) j^*(dx_1) \otimes dy_\nu \,.$$

Then

$$(u_a^* \otimes v^*)(\tilde{\rho}(\omega)) = \sum_{\nu\in T(r,n)} \omega_{1\nu} \circ j \circ u_a d(x_1 \circ j \circ u_a) \otimes d(y_\nu \circ v)$$

$$= \sum_{\nu\in T(r,n)} (\omega_{1\nu} \circ u \circ k) d(x_1 \circ u \circ k) \otimes v^*(dy_\nu).$$

Now

$$u^*(\omega) = \sum_{\kappa+\lambda=t} \sum_{\mu\in T(\kappa,q)} \sum_{\nu\in T(\lambda,n)} \omega_{\mu\nu} \circ u\, d(x_\mu \circ u) \wedge d(y_\nu \circ f \circ u)$$

$$= \sum_{\kappa+\lambda=t} \sum_{\mu\in T(\kappa,q)} \sum_{\nu\in T(\lambda,n)} \omega_{\mu\nu} \circ u\, d(x_\mu \circ u) \wedge d(y_\nu \circ v \circ g).$$

If $\mu \in T(\kappa,q)$ with $\kappa < q$, then $\rho(d(x_\mu \circ u) \wedge d(y_\nu \circ v \circ g)) = 0$. If $\kappa = q$, hence $\mu = 1$, then

$$\rho(d(x_1 \circ u) \wedge d(y_\nu \circ g \circ v)) = d(x_1 \circ u \circ k) \otimes d(y_\nu \circ v).$$

Hence

$$\rho(u^*(\omega)) = \sum_{\nu\in T(r,s)} (\omega_{1\nu} \circ u \circ k)\, d(x_1 \circ u \circ k) \otimes v^*(dy_\nu)$$

$$= (u_a^* \otimes v^*)(\tilde{\rho}(\omega)).$$

Hence it is proven that diagram (4.17) is commutative.

Now, let $\beta: U_\beta \to U'_\beta$ be a diffeomorphism of an open neighborhood U_β of b onto an open subset U'_β of $\mathbb{R}^n$. Set $\beta = (y_1,\ldots,y_n)$. Then $dy_1,\ldots,dy_n$ is a frame field of T(N) over U_β. Hence

$$\tilde{\rho}(\omega) = \sum_{\mu=1}^{n} \omega_\mu \otimes dy_\mu$$

where ω_μ is a form of degree q on $M_b(f)$, which is integrable over $M_b(f)$ for $\mu = 1,\ldots,n$. Hence $u_a^*\omega_\mu$ is integrable over $P_a(g)$ and

$$\int_{P_a(g)} u_a^*(\omega_\mu) = \int_{M_b(f)} \omega_\mu .$$

Moreover,

$$\rho(u^*(\omega)) = u_a^* \otimes v^*(\tilde{\rho}(\omega)) = \sum_{\mu=1}^{n} u_a^*(\omega_\mu) \otimes v^*(dy_\mu).$$

Therefore, $u^*(\omega)$ is integrable over the fibers of g at a and

$$g_*(u^*(\omega)) = \int_{P_a(g)} \rho(u^*(\omega)) = \sum_{\mu=1}^{n} \left(\int_{P_a(g)} u_a^*(\omega_\mu)\right) v^*(dy_\mu)$$

$$= \sum_{\mu=1}^{n} \left(\int_{M_b(f)} \omega_\mu\right) v^*(dy_\mu)$$

$$= v^*\left(\sum_{\mu=1}^{n} \left(\int_{M_b(f)} \omega_\mu\right) dy_\mu\right) = v^* f_*(\omega)$$

at a; q.e.d.

The assumptions of Theorem A II 4.16 shall be studied, when $f: M \to N$ is a fiber bundle with standard fiber F. Here F is a manifold of dimension $q = m - n$. An open covering $\{W_i\}_{i \in I}$ of N by open sets W_j and a family $\{\chi_i\}_{i \in I}$ of diffeomorphisms $\chi_i: f^{-1}(W_j) \to F \times W_i$ exists such that $\sigma_i \circ \chi_i = f$ where σ_i: $F \times W_i \to W_i$ and $\tau_i: F \times W_i \to F$ are the projections. Each fiber $F_a = f^{-1}(a)$ is a manifold such that

$$\chi_{ia} = \sigma_i \circ \chi_i \circ j_a: F_a \to F$$

is a diffeomorphism for $a \in W_i$ where $j_a: F_a \to M$ is the inclusion map. As a fiber of the regular map $f^{-1}(a) = M_a$ is also a manifold. Now, it is claimed that these two orineted, differentiable structures of class C^∞ are the same:

<u>Lemma A II 4.17.</u> <u>For $a \in N$ is $M_a = F_a$.</u>

<u>Proof.</u> Take $i \in I$ such that $a \in W_i$. Take $z_0 \in F_a$. Then $\chi_i(z_0) = (x_0, a)$ with $x_0 \in F$. Let $\gamma: U_\gamma \to U'_\gamma$ be a diffeomorphism of an open neighborhood of U_γ of x_0 onto an open subset U'_γ of $\mathbb{R}^q$. Let $\beta: U_\beta \to U'_\beta$ be a diffeomorphism of an open neighborhood U_β of z_0 with $U_\beta \subseteq W_i$ onto an open subset U'_β of $\mathbb{R}^n$. Define

$$U_\alpha = \chi_j^{-1}(U_\gamma \times U_\beta) \qquad U'_\alpha = U'_\gamma \times U'_\beta$$

$$\alpha = (\gamma \times \beta) \circ \chi_j : U_\alpha \to U'_\alpha .$$

Then α is a diffeomorphism. Let $\pi_\alpha \colon U'_\alpha \to U'_\beta$ and $\psi_\alpha \colon U'_\alpha \to U'_\gamma$ be the projections. Then $\pi_\alpha \circ \alpha = \beta \circ \sigma_j \circ \chi_j = \beta \circ f$ and $\psi_\alpha \circ \alpha = \gamma \circ \tau_j \circ \chi_j$. Hence (α,β) is a product representation of f at z_0. Then $\varepsilon_{(\alpha,\beta)} = \gamma \circ \chi_{ja}$ is a diffeomorphism of the open subset $U_\alpha \cap F_a$ of F_a onto U'_γ. Hence $F_a = M_a$, q.e.d.

Again, let $f\colon M \to N$ be a differentiable fiber bundle with standard fiber F. Let $\{W_i\}_{i\in I}$ and $\{\chi_i\}_{i\in I}$ be given as before. Let S be a manifold of dimension s and let $v\colon S \to N$ be a differentiable map. Define

$$P = \{(z,w) \in M \times S \mid f(z) = v(w)\}.$$

Let $u\colon P \to M$ and $g\colon P \to S$ be the projections. For $i \in I$, define $\tilde{W}_i = v^{-1}(W_i)$. Then $\{\tilde{W}_i\}_{i\in I}$ is an open covering of S with $g^{-1}(\tilde{W}_i) = u^{-1}(f^{-1}(W_i))$ for all $i \in I$. Define $\tilde{\chi}_i\colon g^{-1}(\tilde{W}_i) \to F \times \tilde{W}_i$ by $\tilde{\chi}_i(z,w) = (\tau_i \circ \chi_i(z),w)$ for $(z,w) \in g^{-1}(\tilde{W}_i)$. Then $\tilde{\chi}_i$ is a topological map. Let $\tilde{\sigma}_i\colon F \times \tilde{W}_i \to \tilde{W}_i$ and $\tilde{\tau}_i\colon F \times \tilde{W}_i \to F$ be the projections. Then $\tilde{\sigma}_i \circ \tilde{\chi}_i = g$. If $\tilde{W}_i \cap \tilde{W}_j \neq \emptyset$ then

$$\tilde{\chi}_i \circ \tilde{\chi}_j^{-1}(x,w) = (\tau_i \circ \chi_i \circ \chi_j^{-1}(x,v(w)),w)$$

is a diffeomorphism (of class C^∞ and orientation preserving). Hence P has one and only one structure of a manifold (oriented and of class C^∞) such that $g\colon P \to S$ is a differentiable fiber bundle with the associate families $\{\tilde{W}_i\}_{i\in I}$ and $\{\tilde{\chi}_i\}_{i\in I}$. Especially, each fiber $g^{-1}(w)$ for $w \in S$ has an oriented, C^∞-structure which coincides with the structure given to it by g as a regular map. As usually, this manifold is denoted by P_w. Now

$$g^{-1}(w) = \{(z,w) \in M \times S \mid f(z) = v(w)\} = f^{-1}(v(w)) \times \{w\}.$$

The map $u_w: P_w \to M_{v(w)}$ is given by $u_w(z,w) = z$. For $x \in F$

$$\tau_i \circ \chi_i \circ u_w \tilde{\chi}_i^{-1}(x,w) = x$$

is a diffeomorphism form F onto F, namely the identity, if $w \in \tilde{W}_i$ is fixed. Because $\tau_i \circ \chi_i | M_{v(w)}$ and $\tilde{\chi}_i^{-1}(\cdot,w)$ are diffeomorphic on $M_{v(w)}$ respectively P_w, the map $u_w: P_w \to M_{v(w)}$ is a diffeomorphism. Hence the assumptions of Theorem A II 4.16 are satisfied for the pullback $g: P \to S$ of $f: M \to N$ by v, which proves

Theorem A II 4.18. Let M, N, F, and S be manifolds of dimensions m, n, q and s respectively with $q = m - n > 0$. Let $f: M \to N$ be a differential fiber bundle with general fiber F. Let $v: S \to N$ be a differentiable map. Construct the pullback $g: P \to S$ of $f: M \to N$ by v, where $u: P \to M$ is the associated projection with $f \circ u = v \circ g$. Take $a \in S$ and define $b = v(a)$. Let ω be a form of degree $t \geqq q$ on M which is integrable over the fibers of f at b. Then $u^*(\omega)$ is integrable over the fibers of g at a. Moreover

$$g_* u^* \omega = v^* f_* \omega.$$

An important fact is that integration over the fibers commutes with the exterior derivative:

Theorem A II 4.19. Let M and N be manifolds of dimension m and n respectively such that $q = m - n > 0$. Let $f: M \to N$ be a differentiable and regular map. Let ω be a form of class C^1 and degree $p \geqq q$ on M. Let K be the support of ω. Assume that $f|K: K \to N$ is proper. Then

$$(4.18) \qquad df_*\omega = (-1)^q f_* d\omega$$

on N.

Proof. By Lemma A II 4.9 $f_*\omega$ exists everywhere on N and has class C^1. By the same Lemma, $f_*(d\omega)$ exists everywhere on N and is continuous. Let φ be any form of class C^∞ and degree $m - p - 1 - n - (p-q) - 1$ on N which has compact support on N. Because $f|K$ is proper, $\omega \wedge f^*\varphi$ has compact support on M. Also, $f_*(d\omega) \wedge \varphi$ has compact support on N. Therefore

$$\begin{aligned} \int_N (f_* d\omega) \wedge \varphi &= \int_N f_*(d\omega \wedge f^*\varphi) = \int_M d\omega \wedge f^*\varphi \\ &= \int_M d(\omega \wedge f^*\varphi) + (-1)^{p-1}\int_M \omega \wedge f^*(d\varphi) \\ &= (-1)^{p-1}\int_N f_*\omega \wedge d\varphi \\ &= (-1)^{q+1}\int_N d(f_*\omega \wedge \varphi) + (-1)^q\int_N (df_*\omega) \wedge \varphi \\ &= (-1)^q\int_N (df_*\omega) \wedge \varphi . \end{aligned}$$

Because this holds for all φ, and because $f_*d\omega$ and $df_*\omega$ are continuous (4.18) follows; q.e.d.

§5 The complex analytic case.

Let M be a complex manifold of complex dimension m. Let S(M) be the holomorphic cotangent bundle and $\overline{S}(M)$ the antiholomorphic cotangent bundle of M. The complex manifold M can be considered as a differentiable manifold of real dimension 2m. As such it has a complexified cotangent bundle T(M). Then

$$T(M) = S(M) \oplus \overline{S}(M)$$

$$T^p(M) = \bigoplus_{r+s=p} T^{rs}(M)$$

$$T^{rs}(M) = S^r(M) \wedge S^s(M)$$

where the sections of $T^{rs}(M)$ are the forms of bidegree (r,s). The notation is consistent as $\overline{S}(M)$ is the image of S(M) under the conjugation of T(M). If $\alpha: U_\alpha \to U'_\alpha$ is a biholomorphic map of an open subset U_α of M onto an open subset U'_α of $\mathbb{C}^m$, identify $z = (z_1, \ldots, z_m) \in \mathbb{C}^m$ with $x = (x_1, \ldots, x_{2m}) \in \mathbb{R}^{2m}$ by $z_\mu = x_{2\mu-1} + ix_{2\mu}$ for $\mu = 1, \ldots, m$. Then α becomes a diffeomorphism belonging to the differential structure of M. Moreover, $dz_1, \ldots, dz_m$ is a frame field of S(M) over U_α, and $d\overline{z}_1 \ldots, d\overline{z}_m$ is a frame field of $\overline{S}(M)$ over U_α.

Let f: M → N be a holomorphic map into a complex manifold of complex dimension n. Then f can be considered as a differentiable map and the concepts "smooth" and "regular" apply. Moreover, the maps $\tilde{f}$ and $\hat{f}$ agree with the splitting:

$$\tilde{f}: f^*(S(N)) \to S(N)$$

$$\hat{f}: f^*(S(N)) \to S(M).$$

Moreover, the restriction $\hat{f}_x$ to the fibers over x, is injective (surjective) if and only if f is regular (smooth) at x, which is the case, if and only if

$$f_x^* = \hat{f}_x \circ \tilde{f}_x^{-1}: S_{f(x)}(N) \to S_x(M)$$

is injective (surjective).

The couple (α, β) is said to be a <u>holomorphic product representation</u> of f if and only if

1) The maps $\alpha: U_\alpha \to U'_\alpha$ and $\beta: U_\beta \to U'_\beta$ are biholomorphic maps, where $U_\alpha \subseteq M$ and $U'_\alpha \subseteq \mathbb{C}^m$ and $U_\beta \subseteq N$ and $U'_\beta \subseteq \mathbb{C}^n$ are open.

2) Open subsets U''_α of $\mathbb{C}^q$ and U'''_α of U'_β exist such that $U'_\alpha = U''_\alpha \times U'''_\alpha$ when $\pi_\alpha: U'_\alpha \to U'''_\alpha$ and $\psi_\alpha: U'_\alpha \to U''_\alpha$ are the projections.

3) $$\pi_\alpha \circ \alpha = \beta \circ f.$$

Obviously, a holomorphic product representation is a product representation. The map f is regular at $a \in M$ if and only if a holomorphic product representation of f at a (i.e., $a \in U_\alpha$) exists.

Take $b \in U_\beta$ and let $j\colon f^{-1}(b) \to M$ be the inclusion. Then $f^{-1}(b)$ is an analytic subset of M and as such, $f^{-1}(b)$ has a natural complex structure as a complex space. The inclusion map j is holomorphic, j is smooth exactly at the simple points of $f^{-1}(b)$. If $a \in f^{-1}(b)$ is a regular point of f, then j is smooth at a.

Let (α,β) be a holomorphic product representation of f at $a \in f^{-1}(b)$. Then $a \in M_b(f)$. Moreover,

$$\varepsilon = \psi_\alpha \circ \alpha \circ j\colon f^{-1}(b) \cap U_\alpha \to U''_\alpha$$

if biholomorphic. Therefore, the differentiable structure of $M_b(f)$ agrees with the complex structure of $f^{-1}(b)$. Hence $M_b(f)$ is an open complex submanifold of $f^{-1}(b)$, contained in the set of simple points of $f^{-1}(b)$. Hence, in the complex analytic case, one is relieved of the worry to find the correct orientation on the fibers of f. The results of §4 carry over to the complex analytic case. Because the real fiber dimension of f is always even, the identities of Lemma A II 4.6 and (4.18) read

$$f_*(\omega \wedge f^*\psi) = f_*\omega \wedge \psi \tag{5.1}$$

$$f_*(f^*\psi \wedge \omega) = \psi \wedge f_*\omega \tag{5.2}$$

$$df_*\omega = f_*d\omega \tag{5.3}$$

<u>Proposition A II 5.1.</u> <u>Let M and N be complex manifolds with complex dimensions m and n respectively. Suppose that $q = m - n > 0$. Let</u>

<u>$f\colon M \to N$ be a holomorphic map. Let ω be a form of bidegree (r,s) on M with $r + s \geqq 2q$. Assume that ω is integrable over the fibers of f at $b \in N$. Then $f_*\omega = 0$ if $r < q$ or $s < q$. If $r \geqq q$ and $s \geqq q$, then $f_*\omega$ has bidegree $(r-q, s-q)$.</u>

<u>Proof.</u> Take $a \in M_b(f)$. Let (α,β) be a holomorphic product representation of f at a. Let $\alpha = (z_1,\ldots,z_m)$ and $\beta = (\omega_1,\ldots,\omega_n)$. Then $z_{q+\mu} = \omega_\mu \circ f$ for $\mu = 1,\ldots,n$. Define $t = r - q$ and $u = s - q$. Define

$$\Lambda = \bigcup_{\kappa+\lambda=n} \bigcup_{\varphi+\psi=s} T(\kappa,q) \times T(\lambda,n) \times T(\varphi,q) \times T(\psi,n)$$

Then

$$\omega = \sum_{(\mu,\nu,\eta,\zeta)\in\Lambda} \omega_{\mu\nu\eta\zeta}\, dz_\mu \wedge d\bar{z}_\eta \wedge f^*(d\omega_\nu \wedge d\bar{\omega}_\zeta).$$

Let $j\colon M_b(f) \to M$ be the inclusion. Let i be the sole element of $T(q,q)$. If $t < 0$ or $u < 0$, then $\rho_b(\omega) = 0$. If $t \geqq 0$ and $u \geqq 0$, then

$$\rho_b(\omega) = \sum_{\nu\in T(t,n)} \sum_{\zeta\in T(u,n)} (\omega_{i\nu i\zeta} \circ j)\, j^*(dz_i \wedge d\bar{z}_i) \otimes (d\omega_\nu \wedge d\bar{\omega}_\zeta)$$

Hence $\rho_b(\omega)$ is a section over $M_b(f)$ in

$$T^{2q}(M) \otimes S_b(N)^t \wedge \bar{S}_b(N)^u$$

Therefore $f_*\omega$ has bidegree $(t,u) = (r-q,s-q)$, respectively is zero, if $r < q$ or $s < q$; q.e.d.

Let ω be a form of class C^1 and bidegree (r,s). Then $d\omega = \partial\omega + \bar{\partial}\omega$ where $\partial\omega$ has bidegree $(r+1,s)$ and $\bar{\partial}\omega$ has bidegree $(r,s+1)$. This defines ∂ and $\bar{\partial}$ uniquely such that $d = \partial + \bar{\partial}$. Define

$$d^{\perp} = i(\partial-\bar{\partial}) = -d^c$$

Theorem A II 5.2. Let M and N be complex manifolds of complex manifolds m and n respectively. Assume that $m - n = q > 0$. Let $f\colon M \to N$ be a regular holomorphic map. Let ω be a form of class C^1 and of degree $p \geqq 2q$ on M. Let K be the support of ω. Suppose that $f|K\colon K \to N$ is proper. Then

$$df_*\omega = f_*d\omega \tag{5.4}$$

$$\partial f_*\omega = f_*\partial\omega \tag{5.5}$$

$$\bar{\partial} f_*\omega = f_*\bar{\partial}\omega \tag{5.6}$$

$$d^{\perp}f_*\omega = f_*d^{\perp}\omega \tag{5.7}$$

Proof. (5.4) is correct by Theorem A II 4.18. It is enough to prove the identities for forms with bidegree. Splitting (5.4) by bidegree, if ω has a bidegree, gives (5.5) and (5.6), which implies (5.7); q.e.d.

Let M be a complex manifold of complex dimension m. Let ω be a form of bidegree (p,p) on M. Take $a \in M$. Let $\mathcal{L}_a(p)$ be the set of all smooth, p-dimensional complex submanifolds L of M with $a \in L$. Let $j_L: L \to M$ be the inclusion map. Then $j_L^*(\omega)$ is a form of degree 2p on L. The form ω is said to be non-negative (positive) at a, if and only if $j_L^*(\omega)(a) \geqq 0$ (resp > 0) for all $L \in \mathcal{L}_a(p)$. Write $\omega \geqq 0$ at a (resp $\omega > 0$). If $\omega \geqq 0$ and $\psi \geqq 0$ at a and if ψ has bidegree (1,1) then $\omega \wedge \psi \geqq 0$ at a. If ω has bidegree (p,0), then $(i)^{(p^2)}\omega \wedge \overline{\omega} \geqq 0$ on M.

Lemma A II 5.3. Let M be a complex manifold of complex dimension m. Let ω be a form of bidegree (p,p) on M. Take $a \in M$. Then $\omega \geqq 0$ at a, if and only if[44])

$$i^q\omega \wedge \alpha_1 \wedge \overline{\alpha}_1 \wedge \cdots \wedge \alpha_q \wedge \overline{\alpha}_q \geqq 0 \text{ at } a \tag{5.8}$$

for every q-tuple $\alpha_1,\ldots,\alpha_q$ of forms of bidegree (1,0) on M where $q = m - p$.

Proof. If $\omega \geqq 0$, then $i d\alpha_\mu \wedge d\alpha_\mu \geqq 0$ implies that the form (5.8) is nonnegative.

Suppose that (5.8) is always nonnegative at a. Take $L \in \mathcal{L}_a(p)$. Then a biholomorphic map $\alpha: U_\alpha \to U'_\alpha$ of an open neighborhood U_α of a onto an open neighborhood U'_α of $0 \in \mathbb{C}^m$ with $\alpha(a) = 0$ exists such that $U'_\alpha = U''_\alpha \times U'''_\alpha$ and $\psi \circ \alpha \circ j_L: U_\alpha \cap L \to U''_\alpha$ is biholomorphic. Here $U''_\alpha \subseteq \mathbb{C}^p$ and $U'''_\alpha \subseteq \mathbb{C}^q$ are open and $\psi: U'_\alpha \to U''_\alpha$ is the projection. Moreover, if $\alpha = (z_1,\ldots,z_m)$, then

$$L = \{x \in U_\alpha | z_{p+1}(x) = \ldots = z_m(x) = 0\}$$

such a local coordinate system α exists. Then

$$\omega = (\tfrac{i}{2})^p \sum_{\mu,\nu \in T(p,m)} \omega_{\mu\nu}\, dz_\mu \wedge d\bar{z}_\nu$$

If κ is the only element of $T(p,p)$ then

$$j_L^*(\omega) = (\omega_{\kappa\kappa} \circ j)(\tfrac{i}{2})^p j^*(dz_\kappa \wedge d\bar{z}_\kappa).$$

Now

$$0 \leqq (\tfrac{i}{2})^q \omega \wedge dz_{p+1} \wedge d\bar{z}_{p+1} \wedge \cdots \wedge dz_m \wedge d\bar{z}_m$$

$$= \omega_{\kappa\kappa}(\tfrac{i}{2})^m dz_1 \wedge d\bar{z}_1 \wedge \cdots \wedge dz_m \wedge d\bar{z}_m$$

Hence $\omega_{\kappa\kappa} \geqq 0$ at a. Hence $j^*(\omega) \geqq 0$ at a, q.e.d.

<u>Theorem A II 5.4.</u> <u>Let M and N be complex manifolds of complex dimensions m and n respectively with $m - n = q > 0$. Let $f: M \to N$ be a holomorphic map. Let ω be a form of bidegree (p,p) with $p \geqq q$. Suppose that ω is integrable over the fibers of af at $a \in N$. Suppose that $\omega \geqq 0$ at every $z \in M_a(f)$. Then $f_*\omega \geqq 0$ at a.</u>

Proof. Replacing the value of ω by zero outside $M_a(f)$, it can be assumed that $\omega \geqq 0$ on M. Define $s = m - p$. Let $\alpha_1, \ldots, \alpha_s$ be forms of bidegree (1,0) on N. Then

$$\psi = i^s \omega \wedge f^*(\alpha_1) \wedge \overline{f^*(\alpha_1)} \wedge \ldots \wedge f^*(\alpha_p) \wedge \overline{f^*(\alpha_p)}$$

is nonnegative at a. According to Lemma A II 4.5, Lemma A II 4.6, and Lemma A II 4.10b

$$0 \leqq f_*\psi = i^s (f_*\omega) \wedge \alpha_1 \wedge \overline{\alpha_1} \wedge \ldots \wedge \alpha_s \wedge \overline{\alpha_s}$$

at a, which implies $f_*\omega \geqq 0$ at a; q.e.d.

§6 The integral average.

At first consider the case of real manifolds. Let F and N be manifolds of dimension q and n respectively. Let $\tau: F \times N \to N$ and $\pi: F \times N \to F$ be the projections. Define $M = F \times N$.

Let T(M), T(F) and T(N) be the complexified cotangent bundles. The pullbacks $T_F = \pi^*(T(F))$ and $T_N = \tau^*(T(N))$ can be considered as subbundles of T(M), such that the injections $\tilde{\tau}$ and $\tilde{\pi}$ become inclusion maps. Then

(6.1) $$T(M) = T_F \oplus T_N$$

(6.2) $$T^p(M) = \bigoplus_{r+s=p} T_F^r \wedge T_N^s .$$

A section in $T_F^r \wedge T_N^s$ is said to be a <u>form of type (r,s)</u>; of course, it has degree r + s. If ω is a form of type (r,s) and class C^1, then $d\omega$ splits into a form $d_F\omega$ of type (r+1,s) and a form $d_N\omega$ of type (r,s+1) such that $d\omega = d_F\omega + d_N\omega$. By linearity, the differential operators d_F and d_N are defined on all forms of class C^1 on M or an open subset of M such that $d = d_F + d_N$.

Let ψ be a form of degree q on F. Let ω be a form of type (0,p) on M. Then the integral average $L_\psi(\omega)$ shall be defined in three equivalent ways:

<u>1. Definition</u>: The form $\pi^*(\psi) \wedge \omega$ has type (q,p) and degree q + p on M. If this form is integrable over the fibers of τ at

$a \in N$, define

$$L_\psi(\omega)(a) = \tau^*(\pi^*(\psi) \wedge \omega)(a) \tag{6.3}$$

as the integral average of ω at a for the weight ψ. Obviously, $L_\psi(\omega)$ has degree p where it exists.

2. Definition: The fiber of f over a is $M_a = F \times \{a\}$. Let $j_a\colon M_a \to M$ be the inclusion map. Then $\pi_a = \pi \circ j_a\colon M_a \to F$ is a diffeomorphism. Now

$$\rho_a(\pi^*(\psi) \wedge \omega) = \pi_a^*(\psi) \otimes \omega|M_a \tag{6.4}$$

$$L_\psi(\omega)(a) = \int_{M_a} \pi_a^*(\psi) \otimes \omega|M_a\ . \tag{6.5}$$

Here $T_N^p|M_a = T_a^p(N) \times M_a$ is the trivial bundle and $\omega|M_a$ is a section in the trivial bundle, which corresponds to the section

$$\omega^a = (\mathrm{Id} \times \pi_a) \circ (\omega|M_a) \circ \pi_a^{-1}\colon F \to T_a^p(N) \times F$$

of the trivial bundle $T_a^p(N) \times F$ over F. Hence (6.5) implies

$$L_\psi(\omega)(a) = \int_F \psi \otimes \omega^a \tag{6.6}$$

which gives the second definition.

3. Definition: Take $a \in N$. Let $\beta: U_\beta \to U'_\beta$ be a diffeomorphism of an open neighborhood U_β of a onto an open subset U'_β of R^n. Set $\beta = (y_1,\ldots,y_n)$. Then

(6.7) $$\omega = \sum_{\mu \in T(p,n)} \omega_\mu \tau^*(dy_\mu)$$

on $F \times U_\beta$ where ω_μ are functions on $F \times U_\beta$. Moreover

$$\rho_a(\pi^*(\psi) \wedge \omega) = \sum_{\mu \in T(p,n)} (\omega_\mu \circ j_a)\pi_a^*(\psi) \otimes dy_\mu$$

Hence

(6.8) $$L_\psi(\omega)(a) = \sum_{\mu \in T(p,n)} \left(\int_{x \in F} \omega_\mu(x,a)\psi(x)\right) dy_\mu$$

which is the third definition[45].

Appendix A II §1 and §4 or direct verification by (6.8) imply:

Lemma A II 6.1. If $L_\psi(\omega)$, $L_\varphi(\omega)$ and $L_\psi(\chi)$ exist, at $a \in N$, then $L_{\psi+\varphi}(\omega)$ and $L_\psi(\omega+\chi)$ exist at a with

$$L_{\psi+\varphi}(\omega) = L_\psi(\omega) + L_\varphi(\omega)$$

$$L_\psi(\omega+\chi) = L_\psi(\omega) + L_\psi(\chi).$$

Lemma A II 6.2. If $L_\psi(\omega)$ exists, so does $L_{\overline{\psi}}(\overline{\omega}) = \overline{L_\psi(\omega)}$.

Lemma A II 6.3. If $L_\psi(\omega)$ exists and if χ is a form on N, then $L_\psi(\omega \wedge \tau^*\chi)$ exists with

$$L_\psi(\omega \wedge \tau^*\chi) = L_\psi(\omega) \wedge \chi .$$

Lemma A II 6.4. Suppose that ω is a form of class C^k and type $(0,p)$ on F x N, where $\omega \leqq k \leqq \infty$. Suppose that ψ is a form of degree q on F which is integrable over F and has compact support K in F. Then $L_\psi(\omega)$ exists and is of class C^k on N. If $k \geqq 1$, then

$$dL_\psi(\omega) = L_\psi(d_N\omega).$$

Proof. Represent ω by (6.7). Then (6.8) implies

$$(6.9) \qquad L_\psi(\omega)(z) = \sum_{\mu\in T(p,n)} \left(\int_{x\in K} \omega_\mu(x,z)\psi(x)\right) dy_\mu$$

for $z \in U_\beta$ where ω_μ are of class C^k on F x U_β. Hence $L_\psi(\omega)$ is of class C^k on F x U_β and consequently on F x N. If $k \geqq 1$, then (6.9) implies

$$dL_\psi(\omega)(z) = \sum_{\lambda=1}^{n} \sum_{\mu\in T(p,n)} \int_{x\in K} \left(\frac{\partial}{\partial y_\lambda} \omega_\mu(x,z)\right)\psi(x)\, dy_\lambda \wedge dy_\mu$$

for $z \in U_\beta$. Now

$$d_N\omega = \sum_{\lambda=1}^{n} \sum_{\mu \in T(p,n)} \frac{\partial}{\partial y_\lambda} \omega_\mu(x,z)\, dy_\lambda \wedge dy_\mu$$

at $(x,z) \in F \times U_\lambda$. Therefore, $dL_\psi(\omega) = L_\psi(d_N\omega)$; q.e.d.

Lemma A II 6.5. Let F, N and S be manifolds of dimension q, n and s respectively. Let $\tau: F \times N \to N$ and $\sigma: F \times S \to S$ be the projections. Let $v: S \to N$ be a differentiable map. Define $u: F \times S \to F \times N$ by $u(x,z) = (x,v(z))$. Let ψ be a form of degree q on F. Let ω be a form of type (0,p) on $F \times N$. Then $u^*\omega$ is a form of type (0,p) on $F \times S$. Take $a \in S$ and define $b = v(a)$. If $L_\psi(\omega)$ exists at b, then $L_\psi(u^*\omega)$ exists at a with

$$L_\psi(u^*\omega) = v^*(L_\psi(\omega)).$$

Proof. Theorem A II 4.18.

Lemma A II 6.6. Let F and N manifolds of dimension q and n respectively. Let $\tau: F \times N \to N$ be the projection. For $x \in F$, define $i_x: N \to F \times N$ by $i_x(y) = (x,y)$ for $y \in N$. Let ψ be a measurable form of degree q on F. Let ω be a measurable form of type (0,p) on $F \times N$ with $0 \leq p \leq n$. Let χ be a measurable form of degree n - p on N. Suppose that $i^*_x(\omega) \wedge \chi$ is integrable over N for almost every $x \in F$. Suppose that

$$\left(\int_N |i_x^*\omega \wedge \chi|\right) |\psi(x)|$$

is integrable over F. Suppose that $L_\psi(\omega)$ exists almost everywhere on N.

Then $L_\psi(\omega) \wedge \chi$ is integrable over N with

$$\int_N L_\psi(\omega) \wedge \chi = \int_F (\int_N i_x^* \omega \wedge \chi)\psi .$$

Proof. As before, let π: F x N $\rightarrow$ F be the projection and define M = F x N. Then dim M = m = q + n. The form $\xi = \pi^*(\psi) \wedge \omega \wedge \tau^*(\chi)$ on M has degree m and is measurable. Take a positive form v of degree m and class C^∞ on M. Take any compact subset K of M. For every k $\in$ N, define

$$K_+(k) = \{x \in K | 0 \leqq \xi(x) \leqq kv(x)\}$$

$$K_-(k) = \{x \in K | 0 > \xi(x) \geqq -kv(x)\}$$

$$K_+ = \{x \in K | \xi(x) \geqq 0\}$$

$$K_- = \{x \in K | \xi(x) < 0\} .$$

Let λ_k^+ and λ_k^- be the characteristic functions of $K_+(k)$ respectively $K_-(k)$. Then $\lambda_k^+\xi$ and $\lambda_k^-\xi$ are integrable over M. Let -M be M with the opposite orientation. Now, integration over the fibers of π shall be used. However, the general fiber N of π is the second factor in F x N. Hence the orientation of M has to be changed to $(-1)^{qn}$M. Therefore,

$$0 \leq \int_M \lambda_k^+ \xi = (-1)^{qn} \int_F \pi_*(\lambda_k^+ \pi^* \psi \wedge \omega \wedge \tau^*(\chi))$$

$$= \int_{x \in F} \psi(x) \int_{y \in N} \lambda_k^+(x,y) i_x^* \omega \wedge \chi$$

$$\leq \int_{x \in F} (|\psi(x)| \int_N |i_x^* \omega \wedge \chi|) = C < \infty$$

Similarly, $0 \geq \int_M \lambda_k^- \xi \geq -C$. Observe that C does not depend on K and k. Therefore,

$$C \geq \lim_{k \to \infty} \lambda_k^+ \xi = \int_{K_+} \xi \geq 0$$

$$-C \leq \lim_{k \to \infty} \int_M \lambda_k^- \xi = \int_K \xi \leq 0$$

$$0 \leq \int_K |\xi| = \int_{K_+} \xi - \int_{K-} \xi \leq 2C$$

exist for all compact subsets K. Hence $|\xi|$ is integrable over M. Therefore ξ is integrable over M. Theorem A II 4.11 implies

$$\int_M \xi = (-1)^{qn} \int_F \pi_*(\pi^* \psi \wedge \omega \wedge \tau^*(\chi))$$

$$= \int_{x \in F} (\int_N i_x^*(\omega) \wedge \chi) \psi(x)$$

$$\int_M \xi = \int_N \tau_*(\pi^* \psi \wedge \omega \wedge \tau^* \chi)$$

$$= \int_N \tau_*(\pi^* \psi \wedge \omega) \wedge \chi$$

$$= \int_N L_\psi(\omega) \wedge \chi \qquad \text{q.e.d.}$$

Now, consider the case where N is a complex manifold of complex dimension n. Let $S = S(N)$ be the holomorphic cotangent bundle and $\overline{S}$ its conjugate. The pullbacks to $F \times N$ are $S_N = \tau^*(S)$ and $\overline{S}_N = \tau^*(\overline{S})$. Then

$$T_N = S_N \oplus \overline{S}_N$$

$$T_N^p = \bigoplus_{r+s=p} T_N^{r,s}$$

$$T_N^{r,s} = S_N^r \wedge \overline{S}_N^s$$

A section in $T_F^k \wedge T_N^{r,s}$ is said to be a form of type (k,r,s); of course, it has degree $k + r + s$. If ω is a form of type of class C^1 and (k,r,s), then $d_N\omega$ splits into a form $\partial_N\omega$ of type $(k,r+1,s)$ and $\overline{\partial}_N\omega$ of type $(k,r,s+1)$. By linearity, the differential operators ∂_N and $\overline{\partial}_N$ are defined on all forms of class C^1 on M or an open subset of M such that $d_N = \partial_N + \overline{\partial}_N$. Define $d_N^{\perp} = i(\partial_N - \overline{\partial}_N)$.

Lemma A II 6.7. Let F be a manifold of dimension q. Let ψ be a form of degree q on F. Let N be a complex manifold of complex dimension n. Let ω be a form of type $(0,r,s)$ on $F \times N$. Take $a \in N$. Suppose that $L_\psi(\omega)$ exists at a. Then $L_\psi(\omega)$ has bidegree (r,s) at a.

Proof. Let $\beta\colon U_\beta \to U'_\beta$ be a biholomorphic map of an open neighborhood U_β of a onto an open subset U'_β of $\mathbb{C}^n$. Set $\beta = (z_1,\ldots,z_n)$. Let $\tau\colon F \times N \to N$ be the projection. Then

$$\omega = \sum_{\mu \in T(r,n)} \sum_{\nu \in T(s,n)} \omega_{\mu\nu} \tau^* dz_\mu \wedge \tau^* d\overline{z}_\nu$$

on $F \times U_\beta$ where $\omega_{\mu\nu}$ are functions on $F \times U_\beta$. By (6.8)

$$L_\psi(\omega)(a) = \sum_{\mu \in T(r,n)} \sum_{\nu \in T(s,n)} \left(\int_{x \in F} \omega_{\mu\nu}(x,a)\psi(x)\right) dz_\mu \wedge d\overline{z}_\nu$$

Hence, $L_\psi(\omega)(a)$ has bidegree (r,s); q.e.d.

Degree comparison and Lemma A II 6.4 imply

<u>Lemma A II 6.7.</u> <u>Let F be a manifold of dimension q. Let ψ be a form of degree q on F which has compact support and which is integrable over F. Let N be a complex manifold of complex dimension n. Let ω be a form of class C^1 and type $(0,p)$ on $F \times N$. Then</u>

$$\partial L_\psi(\omega) = L_\psi(\partial_N \omega)$$

$$\overline{\partial} L_\psi(\omega) = L_\psi(\overline{\partial}_N \omega)$$

<u>on N.</u>

<u>Lemma A II 6.8.</u> <u>Let F be a manifold of dimension q. Let ψ be a nonnegative form of degree q on F. Let N be a complex manifold of complex dimension n. For $x \in F$, define $i_x : N \to F \times N$ by $i_x(z) = (x,z)$ for $z \in N$. Let ω be a form of type $(0,p,p)$ on $F \times N$. Suppose that $i_x^*(\omega) \geqq 0$ for every $x \in F$. Take $a \in N$. Suppose that $L_\psi(\omega)$</u>

<u>exists at a. Then $L_\psi(\omega)(a) \geqq 0$.</u>

<u>Proof.</u> At first assume, that $p = n$. Let $\beta: U_\beta \to U'_\beta$ be a holomorphic map of an open neighborhood U_β of a onto an open subset U'_β of $\mathbb{C}^n$. Set $\beta = (z_1,\ldots,z_n)$. Let $\tau: F \times N \to N$ be the projection. Then

$$\omega = (\tfrac{i}{2})^n \quad \omega_0\tau^*(dz_1 \wedge d\bar{z}_1 \wedge \ldots \wedge dz_n \wedge d\bar{z}_n)$$

on $F \times U_\beta$ where ω_0 is a nonnegative function on $F \times U_\beta$. Then

$$L_\psi(\omega)(a) = (\int_{x\in F} \omega_0(x,a)\psi(x))(\tfrac{i}{2})^n dz_1 \wedge d\bar{z}_1 \wedge \ldots \wedge dz_n \wedge d\bar{z}_n.$$

Hence $L_\psi(\omega)(a) \geqq 0$, which proves the case $p = n$.

If $p > n$, then $\omega = 0$. Hence $0 \leqq p < n$ can be assumed. Let L be a p-dimensional, smooth complex submanifold of N such that $a \in L$. Let $v: L \to N$ be the inclusion. Define $u: F \times L \to F \times N$ by $u(x,y) = (x,v(y))$. Let $\sigma: F \times L \to N$ and $\tau: F \times N \to N$ be the projections. Then $\tau \circ u = v \circ \sigma$. For $x \in F$, define $j_x: L \to F \times L$ by $j_x(y) = (x,y)$. Then $u \circ j_x = i_x \circ v$. Lemma A II 6.5 implies

$$v^*(L_\psi(\omega))(a) = L_\psi(u^*\omega)(a).$$

where $u^*\omega$ has type $(0,p,p)$ and where L has complex dimension p. For $x \in F$

$$j_x^*u^*\omega = (u \circ j_x)^*\omega = (i_x \circ v)^*\omega = v^*i_x^*\omega$$

is nonnegative for every $x \in F$. The first part of the proof implies $v^*(L_\psi(\omega))(a) \geq 0$, which - by definition - means $L_\psi(\omega)(a) \geq 0$; q.e.d.

If N is a complex manifold, (or only has even real dimension) then $F \times N$ and $N \times F$ are diffeomorphic having the same orientation. Therefore, the order of the factors does not matter. Of course, the theory also applies if F is a complex manifold in which case d_F splits into $d_F = \partial_F + \bar{\partial}_F$.

Footnotes

1) See Nevanlinna [17] Chapters VI - X.

2) A complex manifold is assumed to be paracompact and to have pure dimension, which is given as the complex dimension. A manifold is supposed to be oriented, of class C^∞, paracompact and of pure dimension.

3) See Example 8.2.

4) Compare Hirschfelder [6] or [7] Definition 6.1.

5) Let $f: M \to N$ be a holomorphic map of a m-dimensional complex manifold into an n-dimensional complex manifold N. Then f is regular (smooth) at $x \in M$ if and only if the Jacobian of f at x has rank $n \leqq m$ (resp. rank $m \leqq n$). A submanifold is smooth if and only if its inclusion map is smooth. See also Appendix II § 3 and §5.

6) Hence $f(x_\lambda) \in S_{a_{\lambda\mu}}$

7) This is the original definition of Lelong [12]. See Appendix II Lemma A II 5.3.

8) See [25] Lemma 7.17.

9) [28] Theorem 4.4 and Hirschfelder [7] §3.

10) [28] Proposition 4.3 and Hirschfelder [7] §3.

11) See [23] Satz 4.5. The proof as given here is due to J. Hirschfelder.

12) See [23] Hilfssatz 1 Page 62.

13) The name is due to Hirschfelder [7].

14) For the definition of $L(\overline{G})$ see page 25.

15) See Appendix II Theorem A II 4.16 and Theorem A II 4.18.

16) The definition of v_{n-1} was given on page 30.

17) Also see Proposition 2.4.

18) See Appendix I Definition A I 12.

19) See Weil [31] and deRham [19] for the proofs of the results mentioned here.

20) Compare [23] and Hirschfelder [7] Theorem 7.5.

21) See Miranda [15] and [25] §1.

22) See [23] Satz 7.3.

23) Compare with Theorem 9.5.

24) $\mathbb{P}(V)$ is a symmetric space. Each form of degree 2j which invariant under all isommetries in harmonic, hence a constant multiple of ω_{0j}.

25) See Chern [2] condition 2 in the Theorem on page 537.

26) Compare Hirschfelder [6] and [7] §2.

27) See [28] Lemma 1.1 and Hirschfelder [6] Lemma 2.2.

28) See [28] Lemma 1.2 and Hirschfelder [6] Lemma 2.3.

29) See Hirschfelder [6], Lemma 2.4.

30) See [28] Lemma 1.3 and Hirschfelder [6] Lemma 2.5.

31) See Hirschfelder [6] Lemma 2.6.

32) See [28] Lemma 1.4 and Hirschfelder [6] Lemma 2.7.

33) See [28] Proposition 1.7 and Hirschfelder [6] Lemma 2.8.

34) See [28] Proposition 1.7.

35) See [28] Proposition 2.8.

36) Compare [21] §3.

37) Compare [28] Lemma 1.8.

38) Recall that an integral over a discreet set is a sum.

39) For instance, see [21].

40) Compare [21].

41) Compare [21] §1.

42) Compare [21] Satz 7.

43) Compare [21] page 134 Zusatz 1.

44) Condition (5.8) is the original definition of Lelong [12].

45) Compare [28] page 175.

References

[1] Bott, R. and Chern, S. S.: Hermitian vector bundles and the equidistribution of the zeros of their holomorphic sections. Acta Math. 114 (1965), 71-112.

[2] Chern, S. S.: The integrated form of the first main theorem for complex analytic mappings in several variables. Ann. of Math. (2) 71, (1960), 536-551.

[3] Chern, S. S.: Some formulas related to complex transgression. (To appear in de Rham Festband, 13 pp. of ms.).

[4] Draper, R.: Intersection theory in analytic gcoemtry. Math. Ann. 180 (1969), 165-204.

[5] Hestens, M.R.: Extension of the range of a differentiable function. Duke Math. J. 8 (1941), 183-192.

[6] Hirschfelder, J.: The first main theorem of value distribution in several variables. (Notre Dame Thesis, 93 pp. of ms.)

[7] Hirschfelder, J.: The first main theorem of value distribution in several variables. Invent. Math. 8 (1969), 1-33, (abbreviated version of [6]).

[7a] Hirschfelder, J.: On Wu's form of the first main theorem of value distribution. Proc. Amer. Math. Soc. 23 (1969), 548-554.

[8] Hörmander, L.: An introduction to complex analysis in several variables. Van Nostrand, Princeton 1966, pp. 208.

[9] Kneser, H.: Zur Theorie der gebrochenen Funktionen mehrerer Veränderlichen. Iber. dtsch Math. Ver. 48, (1938) 1-28.

[10] Kodaira, K. and Spencer, D. C.: On deformations of complex analytic structures, III. Stability theorems for complex structures. Ann. of Math (2) 71 (1960), 43-76.

[11] Lelong, P.: Intégration sure une ensemble analytique complexe. Bull. Soc. Math. France. 85 (1957), 328-370.

[12] Lelong, P.: Eléments postifs d'une algèbre extérieur complexe avec involution. Séminaire d'Analyse. Institute H. Poincarè, Paris, 4-e année 1962 Expose n^o1.

[13] Lelong, P.: Fonctions entières (n variables) et fonctions plurisub harmoniques d'ordre fini dans $\mathbb{C}^n$. J. d'Analyse Math 12 (1964), 365-407.

[14] Levine, H.: A theorem on holomorphic mappings into complex projective space. Ann. of Math. (2) 71 (1960), 529-535.

[15] Miranda, C.: Equazioni alle derivate parziali di tipo ellittico. Ergeb.. Math. Neue Folge 2 (1955) 222 pp.

[16] Narasimhan, R.: The Levi problem for complex spaces. Math. Am. 142 (1960/61), 355-365.

[17] Nevalinna, R.: Eindeutige analytische Funktionen. Die Grundl. d. Math. Wiss. XLVC Springer, Berlin-Göttingen Heidelberg 2. ed. 1953, 379 pp.

[18] Remmert, R.: Holomorphe und meromorphe Abbildungen komplexer Räume. Math. Ann. 133 (1957), 338-370.

[19] de Rham, G.: Varietes differentiables, Hermann, Paris 1955, 196 pp.

[20] Sard, A.: The measure of the cirtical values of differentiable maps. Bull. Am. Math. Soc. 48 (1942), 883-890.

[21] Stoll, W.: Mehrfache Integrale auf komplexen Mannigfaltigkeiten. Math. Z. 57 (1952) 116-154.

[22] Stoll, W.: Ganze Funktionen endlichen Ordnung mit gegebenen Nullstellen-flächen. Math. Z. 57 (1952/53), 211-237.

[23] Stoll, W.: Die beiden Hauptsatze der Wertverteilungs theorie bei Funktionen melinener komplexen Veränderlichen (I). Acta Math. 90 (1953) 1-115.

[24] Stoll, W.: Die beiden Hauptsatze der Wertverteilungstheorie bei Funktionen mehrerer komplexen Veränderlichen (II). Acta Math 92 (1954), 55-169.

[25] Stoll, W.: The growth of the area of a transcendental analytic set. I Math. Ann. 156 (1964) 47-78 and II Math. Ann. 156 (1964) 144-170.

[26] Stoll, W.: The multiplicity of a holomorphic map. Invent. Math. 2 (1966), 15-58.

[27] Stoll, W.: The continuity of the fiber integral. Math. Z. 95 (1967), 87-138.

[28] Stoll, W.: A general first main theory of value distribution. Acta Math. 118 (1967), 111-191.

[29] Stoll, W.: The fiber integral is constant. Math. Z. 104 (1968), 65-73.

[30] Stoll, W.: About the value distribution of holomorphic maps into projective space. Acta Math 123 (1969), 83-114.

[31] Weil, A.: Introduction a l'étude des variétés Kähleriennes. Hermann, Paris, 1968, 175 pp.

[32] Weyl, H. and Weyl, J.: Meromorphic functions and analytic curves. Annals of Math. Studies 12, Princeton University Press, 1943, 269 pp.

[33] Wu, H.: Remarks on the first main theorem of equidistribution theory I. J. Diff. Geom. 2 (1968) 197-202; II J. Diff. Geom. 2 (1968), 369-384; III J. Diff. Geom. 3 (1969) 83-94; (IV to appear in J. Diff. Geom.).

[34] Wu, H.: Mappings of Riemann surfaces (Nevanlinna theory) Proc. Symp. Pure. Math. XI "Entire functions and related topics of Analysis". Amer. Math. Soc. (1968), 480-532.

[35] Wu, H.: The equidistribution theory of holomorphic curves. Berkeley Lecture Notes, 1969, 216 pp.

Index

Offsetdruck: Julius Beltz, Weinheim/Bergstr.

Lecture Notes in Mathematics

Bisher erschienen/Already published

Vol. 1: J. Wermer, Seminar über Funktionen-Algebren. IV, 30 Seiten. 1964. DM 3,80 / $ 1.10

Vol. 2: A. Borel, Cohomologie des espaces localement compacts d'après. J. Leray. IV, 93 pages. 1964. DM 9,– / $ 2.60

Vol. 3: J. F. Adams, Stable Homotopy Theory. Third edition. IV, 78 pages. 1969. DM 8,– / $ 2.20

Vol. 4: M. Arkowitz and C. R. Curjel, Groups of Homotopy Classes. 2nd. revised edition. IV, 36 pages. 1967. DM 4,80 / $ 1.40

Vol. 5: J.-P. Serre, Cohomologie Galoisienne. Troisième édition. VIII, 214 pages. 1965. DM 18,– / $ 5.00

Vol. 6: H. Hermes, Term Logic with Choise Operator. III, 55 pages. 1970. DM 6,– / $ 1.70

Vol. 7: Ph. Tondeur, Introduction to Lie Groups and Transformation Groups. Second edition. VIII, 176 pages. 1969. DM 14,– / $ 3.80

Vol. 8: G. Fichera, Linear Elliptic Differential Systems and Eigenvalue Problems. IV, 176 pages. 1965. DM 13,50 / $ 3.80

Vol. 9: P. L. Ivănescu, Pseudo-Boolean Programming and Applications. IV, 50 pages. 1965. DM 4,80 / $ 1.40

Vol. 10: H. Lüneburg, Die Suzukigruppen und ihre Geometrien. VI, 111 Seiten. 1965. DM 8,– / $ 2.20

Vol. 11: J.-P. Serre, Algèbre Locale. Multiplicités. Rédigé par P. Gabriel. Seconde édition. VIII, 192 pages. 1965. DM 12,– / $ 3.30

Vol. 12: A. Dold, Halbexakte Homotopiefunktoren. II, 157 Seiten. 1966. DM 12,– / $ 3.30

Vol. 13: E. Thomas, Seminar on Fiber Spaces. IV, 45 pages. 1966. DM 4,80 / $ 1.40

Vol. 14: H. Werner, Vorlesung über Approximationstheorie. IV, 184 Seiten und 12 Seiten Anhang. 1966. DM 14,– / $ 3.90

Vol. 15: F. Oort, Commutative Group Schemes. VI, 133 pages. 1966. DM 9,80 / $ 2.70

Vol. 16: J. Pfanzagl and W. Pierlo, Compact Systems of Sets. IV, 48 pages. 1966. DM 5,80 / $ 1.60

Vol. 17: C. Müller, Spherical Harmonics. IV, 46 pages. 1966. DM 5,– / $ 1.40

Vol. 18: H.-B. Brinkmann und D. Puppe, Kategorien und Funktoren. XII, 107 Seiten, 1966. DM 8,– / $ 2.20

Vol. 19: G. Stolzenberg, Volumes, Limits and Extensions of Analytic Varieties. IV, 45 pages. 1966. DM 5,40 / $ 1.50

Vol. 20: R. Hartshorne, Residues and Duality. VIII, 423 pages. 1966. DM 20,– / $ 5.50

Vol. 21: Seminar on Complex Multiplication. By A. Borel, S. Chowla, C. S. Herz, K. Iwasawa, J.-P. Serre. IV, 102 pages. 1966. DM 8,– / $ 2.20

Vol. 22: H. Bauer, Harmonische Räume und ihre Potentialtheorie. IV, 175 Seiten. 1966. DM 14,– / $ 3.90

Vol. 23: P. L. Ivănescu and S. Rudeanu, Pseudo-Boolean Methods for Bivalent Programming. 120 pages. 1966. DM 10,– / $ 2.80

Vol. 24: J. Lambek, Completions of Categories. IV, 69 pages. 1966. DM 6,80 / $ 1.90

Vol. 25: R. Narasimhan, Introduction to the Theory of Analytic Spaces. IV, 143 pages. 1966. DM 10,– / $ 2.80

Vol. 26: P.-A. Meyer, Processus de Markov. IV, 190 pages. 1967. DM 15,– / $ 4.20

Vol. 27: H. P. Künzi und S. T. Tan, Lineare Optimierung großer Systeme. VI, 121 Seiten. 1966. DM 12,– / $ 3.30

Vol. 28: P. E. Conner and E. E. Floyd, The Relation of Cobordism to K-Theories. VIII, 112 pages. 1966. DM 9,80 / $ 2.70

Vol. 29: K. Chandrasekharan, Einführung in die Analytische Zahlentheorie. VI, 199 Seiten. 1966. DM 16,80 / $ 4.70

Vol. 30: A. Frölicher and W. Bucher, Calculus in Vector Spaces without Norm. X, 146 pages. 1966. DM 12,– / $ 3.30

Vol. 31: Symposium on Probability Methods in Analysis. Chairman. D. A. Kappos.IV, 329 pages. 1967. DM 20,– / $ 5.50

Vol. 32: M. André, Méthode Simpliciale en Algèbre Homologique et Algèbre Commutative. IV, 122 pages. 1967. DM 12,– / $ 3.30

Vol. 33: G. I. Targonski, Seminar on Functional Operators and Equations. IV, 110 pages. 1967. DM 10,– / $ 2.80

Vol. 34: G. E. Bredon, Equivariant Cohomology Theories. VI, 64 pages. 1967. DM 6,80 / $ 1.90

Vol. 35: N. P. Bhatia and G. P. Szegö, Dynamical Systems. Stability Theory and Applications. VI, 416 pages. 1967. DM 24,– / $ 6.60

Vol. 36: A. Borel, Topics in the Homology Theory of Fibre Bundles. VI, 95 pages. 1967. DM 9,– / $ 2.50

Vol. 37: R. B. Jensen, Modelle der Mengenlehre. X, 176 Seiten. 1967. DM 14,– / $ 3.90

Vol. 38: R. Berger, R. Kiehl, E. Kunz und H.-J. Nastold, Differentialrechnung in der analytischen Geometrie IV, 134 Seiten. 1967 DM 12,– / $ 3.30

Vol. 39: Séminaire de Probabilités I. II, 189 pages. 1967. DM 14,– / $ 3.90

Vol. 40: J. Tits, Tabellen zu den einfachen Lie Gruppen und ihren Darstellungen. VI, 53 Seiten. 1967. DM 6,80 / $ 1.90

Vol. 41: A. Grothendieck, Local Cohomology. VI, 106 pages. 1967. DM 10,– / $ 2.80

Vol. 42: J. F. Berglund and K. H. Hofmann, Compact Semitopological Semigroups and Weakly Almost Periodic Functions. VI, 160 pages. 1967. DM 12,– / $ 3.30

Vol. 43: D. G. Quillen, Homotopical Algebra. VI, 157 pages. 1967. DM 14,– / $ 3.90

Vol. 44: K. Urbanik, Lectures on Prediction Theory. IV, 50 pages. 1967. DM 5,80 / $ 1.60

Vol. 45: A. Wilansky, Topics in Functional Analysis. VI, 102 pages. 1967. DM 9,60 / $ 2.70

Vol. 46: P. E. Conner, Seminar on Periodic Maps.IV, 116 pages. 1967. DM 10,60 / $ 3.00

Vol. 47: Reports of the Midwest Category Seminar I. IV, 181 pages. 1967. DM 14,80 / $ 4.10

Vol. 48: G. de Rham, S. Maumary et M. A. Kervaire, Torsion et Type Simple d'Homotopie. IV, 101 pages. 1967. DM 9,60 / $ 2.70

Vol. 49: C. Faith, Lectures on Injective Modules and Quotient Rings. XVI, 140 pages. 1967. DM 12,80 / $ 3.60

Vol. 50: L. Zalcman, Analytic Capacity and Rational Approximation. VI, 155 pages. 1968. DM 13,20 / $ 3.70

Vol. 51: Séminaire de Probabilités II. IV, 199 pages. 1968. DM 14,– / $ 3.90

Vol. 52: D. J Simms, Lie Groups and Quantum Mechanics. IV, 90 pages. 1968. DM 8,– / $ 2.20

Vol. 53: J. Cerf, Sur les difféomorphismes de la sphère de dimension trois (Γ_4= O). XII, 133 pages. 1968. DM 12,– / $ 3.30

Vol. 54: G. Shimura, Automorphic Functions and Number Theory. VI, 69 pages. 1968. DM 8,– / $ 2.20

Vol. 55: D. Gromoll, W. Klingenberg und W. Meyer, Riemannsche Geometrie im Großen. VI, 287 Seiten. 1968. DM 20,– / $ 5.50

Vol. 56: K. Floret und J. Wloka, Einführung in die Theorie der lokalkonvexen Räume. VIII, 194 Seiten. 1968. DM 16,– / $ 4.40

Vol. 57: F. Hirzebruch und K. H. Mayer, O (n)-Mannigfaltigkeiten, exotische Sphären und Singularitäten. IV, 132 Seiten. 1968. DM 10,80 / $ 3.00

Vol. 58: Kuramochi Boundaries of Riemann Surfaces. IV, 102 pages. 1968. DM 9,60 / $ 2.70

Vol. 59: K. Jänich, Differenzierbare G-Mannigfaltigkeiten. VI, 89 Seiten. 1968. DM 8,– / $ 2.20

Vol. 60: Seminar on Differential Equations and Dynamical Systems. Edited by G. S. Jones. VI, 106 pages. 1968. DM 9,60 / $ 2.70

Vol. 61: Reports of the Midwest Category Seminar II. IV, 91 pages. 1968. DM 9,60 / $ 2.70

Vol. 62: Harish-Chandra, Automorphic Forms on Semisimple Lie Groups X, 138 pages. 1968. DM 14,– / $ 3.90

Vol. 63: F. Albrecht, Topics in Control Theory. IV, 65 pages. 1968. DM 6,80 / $ 1.90

Vol. 64: H. Berens, Interpolationsmethoden zur Behandlung von Approximationsprozessen auf Banachräumen. VI, 90 Seiten. 1968. DM 8,– / $ 2.20

Vol. 65: D. Kölzow, Differentiation von Maßen. XII, 102 Seiten. 1968. DM 8,– / $ 2.20

Vol. 66: D. Ferus, Totale Absolutkrümmung in Differentialgeometrie und -topologie. VI, 85 Seiten. 1968. DM 8,– / $ 2.20

Vol. 67: F. Kamber and P. Tondeur, Flat Manifolds. IV, 53 pages. 1968. DM 5,80 / $ 1.60

Vol. 68: N. Boboc et P. Mustaţă, Espaces harmoniques associés aux opérateurs différentiels linéaires du second ordre de type elliptique. VI, 95 pages. 1968. DM 8,60 / $ 2.40

Vol. 69: Seminar über Potentialtheorie. Herausgegeben von H. Bauer. VI, 180 Seiten. 1968. DM 14,80 / $ 4.10

Vol. 70: Proceedings of the Summer School in Logic. Edited by M. H. Löb. IV, 331 pages. 1968. DM 20,– / $ 5.50

Vol. 71: Séminaire Pierre Lelong (Analyse), Année 1967 – 1968. VI, 19 pages. 1968. DM 14,– / $ 3.90

Bitte wenden / Continued

Vol. 72: The Syntax and Semantics of Infinitary Languages. Edited by J. Barwise. IV, 268 pages. 1968. DM18,– / $ 5.00

Vol. 73: P. E. Conner, Lectures on the Action of a Finite Group. IV, 123 pages. 1968. DM 10,– / $ 2.80

Vol. 74: A. Fröhlich, Formal Groups. IV, 140 pages. 1968. DM12,– / $ 3.30

Vol. 75: G. Lumer, Algèbres de fonctions et espaces de Hardy. VI, 80 pages. 1968. DM 8,– / $ 2.20

Vol. 76: R. G. Swan, Algebraic K-Theory. IV, 262 pages. 1968. DM18,– / $ 5.00

Vol. 77: P.-A. Meyer, Processus de Markov: la frontière de Martin. IV, 123 pages. 1968. DM 10,– / $ 2.80

Vol. 78: H. Herrlich, Topologische Reflexionen und Coreflexionen. XVI, 166 Seiten. 1968. DM 12,– / $ 3.30

Vol. 79: A. Grothendieck, Catégories Cofibrées Additives et Complexe Cotangent Relatif. IV, 167 pages. 1968. DM 12,– / $ 3.30

Vol. 80: Seminar on Triples and Categorical Homology Theory. Edited by B. Eckmann. IV, 398 pages. 1969. DM 20,– / $ 5.50

Vol. 81: J.-P. Eckmann et M. Guenin, Méthodes Algébriques en Mécanique Statistique. VI, 131 pages. 1969. DM 12,– / $ 3.30

Vol. 82: J. Wloka, Grundräume und verallgemeinerte Funktionen. VIII, 131 Seiten. 1969. DM 12,– / $ 3.30

Vol. 83: O. Zariski, An Introduction to the Theory of Algebraic Surfaces. IV, 100 pages. 1969. DM 8,– / $ 2.20

Vol. 84: H. Lüneburg, Transitive Erweiterungen endlicher Permutationsgruppen. IV, 119 Seiten. 1969. DM 10,– / $ 2.80

Vol. 85: P. Cartier et D. Foata, Problèmes combinatoires de commutation et réarrangements. IV, 88 pages. 1969. DM 8,– / $ 2.20

Vol. 86: Category Theory, Homology Theory and their Applications I. Edited by P. Hilton. VI, 216 pages. 1969. DM 16,– / $ 4.40

Vol. 87: M. Tierney, Categorical Constructions in Stable Homotopy Theory. IV, 65 pages. 1969. DM 6,– / $ 1.70

Vol. 88: Séminaire de Probabilités III. IV, 229 pages. 1969. DM18,– / $ 5.00

Vol. 89: Probability and Information Theory. Edited by M. Behara, K. Krickeberg and J. Wolfowitz. IV, 256 pages. 1969. DM 18,– / $ 5.00

Vol. 90: N. P. Bhatia and O. Hajek, Local Semi-Dynamical Systems. II, 157 pages. 1969. DM 14,– / $ 3.90

Vol. 91: N. N. Janenko, Die Zwischenschrittmethode zur Lösung mehrdimensionaler Probleme der mathematischen Physik. VIII, 194 Seiten. 1969. DM 16,80 / $ 4.70

Vol. 92: Category Theory, Homology Theory and their Applications II. Edited by P. Hilton. V, 308 pages. 1969. DM 20,– / $ 5.50

Vol. 93: K. R. Parthasarathy, Multipliers on Locally Compact Groups. III, 54 pages. 1969. DM 5,60 / $ 1.60

Vol. 94: M. Machover and J. Hirschfeld, Lectures on Non-Standard Analysis. VI, 79 pages. 1969. DM 6,– / $ 1.70

Vol. 95: A. S. Troelstra, Principles of Intuitionism. II, 111 pages. 1969. DM 10,– / $ 2.80

Vol. 96: H.-B. Brinkmann und D. Puppe, Abelsche und exakte Kategorien, Korrespondenzen. V, 141 Seiten. 1969. DM 10,– / $ 2.80

Vol. 97: S. O. Chase and M. E. Sweedler, Hopf Algebras and Galois theory. II, 133 pages. 1969. DM 10,– / $ 2.80

Vol. 98: M. Heins, Hardy Classes on Riemann Surfaces. III, 106 pages. 1969. DM 10,– / $ 2.80

Vol. 99: Category Theory, Homology Theory and their Applications III. Edited by P. Hilton. IV, 489 pages. 1969. DM 24,– / $ 6.60

Vol. 100: M. Artin and B. Mazur, Etale Homotopy. II, 196 Seiten. 1969. DM 12,– / $ 3.30

Vol. 101: G. P. Szegö et G. Treccani, Semigruppi di Trasformazioni Multivoche. VI, 177 pages. 1969. DM 14,– / $ 3.90

Vol. 102: F. Stummel, Rand- und Eigenwertaufgaben in Sobolewschen Räumen. VIII, 386 Seiten. 1969. DM 20,– / $ 5.50

Vol. 103: Lectures in Modern Analysis and Applications I. Edited by C. T. Taam. VII, 162 pages. 1969. DM 12,– / $ 3.30

Vol. 104: G. H. Pimbley, Jr., Eigenfunction Branches of Nonlinear Operators and their Bifurcations. II, 128 pages. 1969. DM 10,– / $ 2.80

Vol. 105: R. Larsen, The Multiplier Problem. VII, 284 pages. 1969. DM 18,– / $ 5.00

Vol. 106: Reports of the Midwest Category Seminar III. Edited by S. Mac Lane. III, 247 pages. 1969. DM 16,– / $ 4.40

Vol. 107: A. Peyerimhoff, Lectures on Summability. III, 111 pages. 1969. DM 8,– / $ 2.20

Vol. 108: Algebraic K-Theory and its Geometric Applications. Edited by R. M.F. Moss and C. B. Thomas. IV, 86 pages. 1969. DM 6,– / $ 1.70

Vol. 109: Conference on the Numerical Solution of Differential Equations. Edited by J. Ll. Morris. VI, 275 pages. 1969. DM 18,– / $ 5.00

Vol. 110: The Many Facets of Graph Theory. Edited by G. Chartrand and S. F. Kapoor. VIII, 290 pages. 1969. DM 18,– / $ 5.00

Vol. 111: K. H. Mayer, Relationen zwischen charakteristischen Zahlen. III, 99 Seiten. 1969. DM 8,– / $ 2.20

Vol. 112: Colloquium on Methods of Optimization. Edited by N. N. Moiseev. IV, 293 pages. 1970. DM 18,– / $ 5.00

Vol. 113: R. Wille, Kongruenzklassengeometrien. III, 99 Seiten. 1970. DM 8,– / $ 2.20

Vol. 114: H. Jacquet and R. P. Langlands, Automorphic Forms on GL (2). VII, 548 pages. 1970. DM 24,– / $ 6.60

Vol. 115: K. H. Roggenkamp and V. Huber-Dyson, Lattices over Orders I. XIX, 290 pages. 1970. DM 18,– / $ 5.00

Vol. 116: Séminaire Pierre Lelong (Analyse) Année 1969. IV, 195 pages. 1970. DM 14,– / $ 3.90

Vol. 117: Y. Meyer, Nombres de Pisot, Nombres de Salem et Analyse Harmonique. 63 pages. 1970. DM 6,– / $ 1.70

Vol. 118: Proceedings of the 15th Scandinavian Congress, Oslo 1968. Edited by K. E. Aubert and W. Ljunggren. IV, 162 pages. 1970. DM 12,– / $ 3.30

Vol. 119: M. Raynaud, Faisceaux amples sur les schémas en groupes et les espaces homogènes. III, 219 pages. 1970. DM 14,– / $ 3.90

Vol. 120: D. Siefkes, Büchi's Monadic Second Order Successor Arithmetic. XII, 130 Seiten. 1970. DM 12,– / $ 3.30

Vol. 121: H. S. Bear, Lectures on Gleason Parts. III, 47 pages. 1970. DM 6,– / $ 1.70

Vol. 122: H. Zieschang, E. Vogt und H.-D. Coldewey, Flächen und ebene diskontinuierliche Gruppen. VIII, 203 Seiten. 1970. DM 16,– / $ 4.40

Vol. 123: A. V. Jategaonkar, Left Principal Ideal Rings. VI, 145 pages. 1970. DM 12,– / $ 3.30

Vol. 124: Séminare de Probabilités IV. Edited by P. A. Meyer. IV, 282 pages. 1970. DM 20,– / $ 5.50

Vol. 125: Symposium on Automatic Demonstration. V, 310 pages. 1970. DM 20,– / $ 5.50

Vol. 126: P. Schapira, Théorie des Hyperfonctions. XI, 157 pages. 1970. DM 14,– / $ 3.90

Vol. 127: I. Stewart, Lie Algebras. IV, 97 pages. 1970. DM 10,– / $ 2.80

Vol. 128: M. Takesaki, Tomita's Theory of Modular Hilbert Algebras and its Applications. II, 123 pages. 1970. DM 10,– / $ 2.80

Vol. 129: K. H. Hofmann, The Duality of Compact Semigroups and C*-Bigebras. XII, 142 pages. 1970. DM 14,– / $ 3.90

Vol. 130: F. Lorenz, Quadratische Formen über Körpern. II, 77 Seiten. 1970. DM 8,– / $ 2.20

Vol. 131: A Borel et al., Seminar on Algebraic Groups and Related Finite Groups. VII, 321 pages. 1970. DM 22,– / $ 6.10

Vol. 132: Symposium on Optimization. III, 348 pages. 1970. DM 22,– / $ 6.10

Vol. 133: F. Topsøe, Topology and Measure. XIV, 79 pages. 1970. DM 8,– / $ 2.20

Vol. 134: L. Smith, Lectures on the Eilenberg-Moore Spectral Sequence. VII, 142 pages. 1970. DM 14,– / $ 3.90

Vol. 135: W. Stoll, Value Distribution of Holomorphic Maps into Compact Complex Manifolds. II, 267 pages. 1970. DM 18,– / $ 5.00